Flat Earth

Conspiracy Theories About the
Earth's Surface

*(The Flat Earth Trilogy Book of Secrets Discovey
Inside the Earth)*

Doris Carter

Published By **Phil Dawson**

Doris Carter

All Rights Reserved

Flat Earth: Conspiracy Theories About the Earth's Surface (The Flat Earth Trilogy Book of Secrets Discovey Inside the Earth)

ISBN 978-1-77485-719-9

No part of this guidebook shall be reproduced in any form without permission in writing from the publisher except in the case of brief quotations embodied in critical articles or reviews.

Legal & Disclaimer

The information contained in this ebook is not designed to replace or take the place of any form of medicine or professional medical advice. The information in this ebook has been provided for educational & entertainment purposes only.

The information contained in this book has been compiled from sources deemed reliable, and it is accurate to the best of the Author's knowledge; however, the Author cannot guarantee its accuracy and validity and cannot be held liable for any errors or omissions. Changes are periodically made to this book. You must consult your doctor or get professional medical advice before using any of the suggested remedies, techniques, or information in this book.

Upon using the information contained in this book, you agree to hold harmless the Author from and against any damages, costs, and expenses, including any legal fees potentially resulting from the application of any of the information provided by this guide. This disclaimer applies to any damages or injury caused by the use and application, whether directly or indirectly, of any advice or information presented, whether for breach of contract, tort, negligence, personal injury, criminal intent, or under any other cause of action.

You agree to accept all risks of using the information presented inside this book. You need to consult a professional medical practitioner in order to ensure you are both able and healthy enough to participate in this program.

Table Of Contents

Introduction _____ 1

Chapter 1: Ancient Beliefs _____ 4

Chapter 2: Revision Of The Flat Earth Theory_____ 26

Chapter 3: "The Flat Earth Movement In The 20th Century _____ 49

Chapter 4: The Internet _____ 67

Chapter 5: How Do Flat Earthers Think? _____ 80

Chapter 6: Flat Earthers Throughout History _____ 114

Chapter 7: Cameras Don't Lie ___ 144

Chapter 8: The Stars_____ 166

Introduction

Flat Earth

The world is full of mysteries. And even today some of the planet is still unexplored. The depths of oceans as well as the complex and vast cave systems that honeycomb certain regions of the Earth remain largely unexplored. So, it should come as no surprise when you think about this terra inaccessible, people have formulated many theories. The stories of sunken cities and lost civilizations are only a few of the many fanciful theories and they could be considered to be unorthodox in comparison to the notion that the Earth is flat.

Despite this belief being dismissed by scientists for centuries, and despite the fact that volcanology, geology oceanography, and physics have proven that the Earth isn't flat, the notion of hollow Earth is still a topic of fascination and attract enthusiastic and committed followers. This is made more impressive due to the fact that space programs are over 60 years old and that people are able to travel the globe in planes in just a few hours.

When taken as a whole the idea seems absurd, yet it triggers intense emotions in certain people honest people who have pondered for a long time about their convictions. People who believe they know an unspoken truth and the majority of humanity is ignorant and wrong however, this feeling of superiority in mind isn't the only reason for adhering to the radical ideas. There's also the thrill of adventure, the sensation that you are part of a dangerous group who are trying to change the accepted model. It's much more enjoyable, as some feel to live in a world that is full of excitement and mystery, as opposed to a degrading, "rational" world where every aspect can be explained, but there is no solution.

Chapter 1: Ancient Beliefs

The ancients were forgiven to believe that the universe was flat due to the fact that the Earth was a lot bigger for humans to be able to move around it. Even when we are on top of the top of a mountain, the Earth appears flat enough, and the notion of an elongated globe that was spinning around space seemed counterintuitive to people who did not have a clue about Physics or astronomy. So, it's no surprise that throughout time, people believed that they were on the Earth was flat. more than likely that the majority of people didn't give any thought at all to this being consumed by the everyday effort to live. War, disease and gruelling manual labor were a regular part of most people's lives. And they did not have time to focus contemplating the geography of the world.

However, the ancient civilizations had scholars and priests who contemplated on the mystery of the universe and even thought what the form of the Earth. Both Mesopotamians as well as the ancient Egyptians believed that the Earth was flat and floating upon a huge ocean. the Mesopotamians imagined the world as a disk , whereas the Egyptians believed it was a square shape. The early Israelites also saw that the universe as disc floating over an ocean and also believed that the earth was covered by an enveloping dome in which all celestial bodies were encased.

Chinese scholars believed that the earth was square and flat and disputed the nature of the universe. Many believed that it was suspended over the Earth like an umbrella however, others believed that it was a sphere around it. Some people came closer to truth when they claimed that it had no substance whatsoever and that the stars could be seen floating in it. Because of the stifling nature of education there the world was not advancing regarding the idea of a world until contact to European merchants in 17th-century Europe brought the concept of a circular Earth.

It is believed that the Indian civilizations were superior in this regard. Although it is true that the Vedic texts, the primary source of Hindu faith, portray the globe as flat or concave, and disk-shaped and shaped, later on, the field of astronomy developed into a science on the subcontinent. In in the fifth century CE Astronomers had come to the realization that they were aware that the Earth could be viewed as a globe, and several philosophers had incorporated this understanding into their theories. Certain historians have suggested some Indian academics were inspired by modern Greek knowledge, however it is generally believed that they came up with these discoveries independently.

The early Greeks believed that there was a disc-shaped world, with water surrounding it, and Greek philosophers would argue for hours over the precise nature of the universe. The Anaximander (c. 610-546 BCE) believed that the Earth was one of stubby cylinders with an unflat, circular top. The circular cylinder was about 1/3 as tall as it was wide, so one might also imagine it as a large disk. The people lived on the top of this cylinder and it stood on its own and didn't move since it was located at the center of all things, an unusually early description of the notion of a Lagrangian point. In the model of Anaximander skies weren't a solid object or an edifice, but an array of concentric wheels within which celestial bodies were fixed and around the wheels they moved. The wheels were equipped with holes so that people could see the celestial bodies. However the holes on the wheels' rims could be blocked in some instances and this could be the reason for the various patterns associated with those of the Moon as well as eclipses. Because the Earth appears to float in space and this model allows celestial bodies to move under the Earth and provide a clear explanation of why they set and rise. He also considered as the Sun as a gigantic spaceship that was far from the Earth. The model of Anaximander even allowed different worlds to emerge and die, and no the way he came to the conclusion, he was on the right track in assuming that the Earth had a short lifespan.

A rendering of Anaximander's theories of the world.

Anaximenes who lived in Miletus (c. 586-526 BCE) believed "the Earth is flat and is riding on air similar to how the sun and moon and other celestial bodies, all fiery, travel on air due to the flatness of their bodies." He likened these bodies to leaflets floating on the air. It is believed that the Sun as well as the Moon are both flat disks, similar to the Earth and have their faces pointed towards the direction of the planet. They orbit around the Earth but they do not travel under it, but instead are hidden by mountains on the edge of the planet.

Xenophanes who were part of Colophon (c. between 570 and 475 BCE) suggested in the era of Xenophanes that Earth was flat with its top side touching the air , and its lower half going down indefinitely. As with anaximander, he believed that there was the possibility of a universe with infinite possibilities for that were born, and then eventually dying. Incredibly, Xenophanes was also a pioneer proponent of evolution, a scientific concept that was largely rejected by Flat Earthers. He studied fossils and concluded that the only reason for the discovery of fossils of fish and sea creatures high up on the mountain slopes was that the planet was once submerged in water.

The man believed in the fact that life was formed from primordial mud and water.

The early Greeks were able to come up with innovative theories about the origins of Earth however, later another group of earlier Greeks created one of the most significant leaps in the field of science by positing it was possible that Earth was actually a globe. The 6th century BCE two natural philosophers were the first to propose this concept. Parmenides from Elea (c. 515 BCE) was possibly the first to suggest the idea of a globe model, however the more well-known Pythagoras (c. 557-495 BCE) also supported this idea and has been credited as the one who discovered. Their theories were accepted by educated Greeks at the close in the fifth century BCE and, with it the Earth was considered to be the central point of the universe, with it being the center of the universe with the Sun, Moon, planets and stars being orbited around the Earth. The geocentric model was in use for longer than the notion of the Flat Earth and wasn't disproved until Copernicus published his famous research De Revolutionibus Orbium coelestium (On the

The Heavenly Spheres' Revolutions) around 1543. Although many had proposed an heliocentric theory before Copernicus, such as the ancient Greek Astronomer Aristarchus from Samos (c. 308-230 BCE) but it was Copernicus who changed the minds of many and helped push science into a new direction.

The bust is of Pythagoras

The last defender of this Flat Earth model among Greek philosophers was Archelaus (lived during the 5th century BCE). He came up with an interesting variation of the model that suggested it was possible that the Earth was compressed to the middle of the earth as the saucer. This would allow the sun to set and rise at different times based upon where on the Earth you were. The common observation that days at north and south latitudes last longer in summer , and shorter in winter than those on the equator has been an issue for those who support the use who believe in Flat Earth. Flat Earth model.

Archelaus was one of the men who was born to die too late. At the time of his birth there was a consensus to believe that Earth is a globe however, little was known about the fact that. Eratosthenes from Cyrene (c. 276 to 244 BCE) advanced the understanding of humanity one step further when he calculated how wide the Earth. The Greek scholar was a scholar from Egypt and was aware that on the solstice of summer that there was a Sun during noon, was overhead in Syene the modern city of Aswan close to the current boundary with Egypt as well as Sudan. It was widely recognized that the Sun at that time reflected through the frozen bottom of the well, which meant that it was directly over the horizon. Eratosthenes could determine how far from Syene as well as Alexandria in the Mediterranean and assuming that the Earth was perfectly spherical, Eratosthenes discovered that he could calculate its dimensions by taking the angle of the sun in Alexandria in the noon hour of this same date. Because Alexandria was a different place on the planet that the Sun will not be directly overhead. However, by measuring its angle , and then doing some basic maths, he could determine the distance in angular terms between the two cities.

He did this using an instrument called a gnomon. It measured its shadow over the ground. With how long the rod as well as the distance of the shadow the gnomon could be seen in terms of two triangles, and therefore calculate how much angle is created by the sun's radiations. The angle was about seven degrees, which is about 2 per cent of the diameter the circle. In other words, he claimed the diameter of Earth would be 50 times greater than the length between Alexandria to Syene.

The number Eratosthenes had in mind was 252,000 stadia. This was a measurement that was used for different lengths in the past due to the fact that there was no uniform measure. It's not known which stade Eratosthenes used, however their lengths are all well-known and are able to be converted into modern measurements , which give an approximate range of 27400 to 322,500 miles (44,100-52,300 kilometers). The actual number is 24,860 miles. This is based on the fragility of the instruments Eratosthenes had to work as well as the inability to know the exact measurement between Alexandria and Syene and the reality it is the case that Earth is in fact an oblate spheroid, and not as perfectly spherical as it was thought to be, Eratosthenes's measurements were astonishing accomplishments. Even if his measurements were a little off, he was able to prove that the Earth is round.

Following that after that, the Greeks nearly unanimously believed that the universe was circular the world was round, a notion that was accepted (like many the other Greek wisdom) from their admirers as well as conquerors the Romans. One of the only Romans who opposed this idea included Titus Lucretius Carus (99-55 BCE) whom wrote in De rerum natura (On the Nature of Things) that it was absurd to think that people from the opposite side of the globe had a tendency to walk upside down, a belief that was echoed more than 2,000 years later by contemporary Flat Earthers. In this respect, Carus was truly a last defender, as a large portion of the world's ancient people were today firmly in the belief of the round Earth and this belief was passed down to Muslim civilizations as well as the European ancient civilizations.

One thing that muddied the waters for a few people was the advent of Christianity and the fact that the Bible that was believed to be the most authoritative source in all things, includes numerous passages that appear to suggest that the Earth can be flat. For instance in the account of Jesus' Temptation of Christ within Matthew 4:8 in the gospel, it is stated that "The Devil took him to an extremely high mountain and showed him of the nations of all the earth and its beauty." This suggests that from an incredibly high location, one could view the entire globe which isn't feasible if the Earth was an earth. The same story is repeated in Luke 4:5 and a similar story can be located in Daniel 4:10-11. It states, "Before me stood a tree that was in between the country. Its size was immense. The tree grew enormous and strong , and its top was in the sky. It was visible from the ends on globe." This statement suggests that a tall object can be seen from all points in the globe, suggesting that it is that it is a plane, not an actual globe. The statement also states that the Earth is not complete and the sky could be touched, which supports theories like the "enclosed disc and vault" theory.

Certain passages, like Isaiah 11:12 or Revelation 7:1, speak of"the "four points of earth" that those who follow the Flat Earth model have taken to mean that the universe isn't just a plane, but also a rectangle or square.

Job 26:7 is a fascinating section: "He [God] spreads out the northern skies across empty space. He suspends the earth above the ground." This is believed as a sign that there is nothing under the earth and therefore celestial bodies like the Sun as well as the Moon don't travel to the opposite side of the planet when they set.

The First Book of Chronicles 16:30 It says "The world is fixed and is not able to be changed." The same sentence appears in the Psalms 93:1, and has been interpreted by a lot of readers throughout earlier times in the Middle Ages and more modern times to indicate that the Earth cannot be moving around the Sun or bouncing through space at incredible speeds , as astronomers have found.

Although these passages may cause trouble for people who adhered to the model of the world of the Earth but they actually hindered the learning process very little. Classical learning didn't disappear completely during the "Dark The Ages" in Europe and the majority of the limited research within Western Europe during the centuries following the fall of the Roman Empire continued to teach the model of the globe. In the Byzantine Empire also kept this model of the globe.

Some of the most influential thinkers, however, resisted the notion, and one of them is Saint Augustine. In his work The City of God, the influential Church father wrote "But in relation to the myth that there are Antipodes, which means, they are who are on the other end of Earth, in the place where the sun rises as it is set to us, people who have feet in opposition to ours, it is not a fact that can be believed. It isn't proven that this fact has been discovered through historical research or by scientific speculation on the assumption of the fact that Earth is suspended in the concavity of our sky, and that it occupies plenty of space on one side as it does on the opposite side; so the portion below should also be habitable. They do not mention that, despite it being claimed or proven scientifically that the earth is of an spherical and round shape however, it doesn't necessarily mean that the other portion of Earth is devoid of water. Nor when it is bare and waterless, it does not mean that it is populated. Since Scripture is the only source of proof for the authenticity of its historical claims through the fulfillment of its prophecies gives an accurate account of the world; and it's absurd to think that some people could have taken a ship and traveled the entire ocean and traveled from one end of the earth to the other side, and, consequently, even the people of that far region descend from that one man who was the first. So let us look to see to find that city God that dwells within the human race that are said to have been divided into seventy two nations and several languages. It lasted until the flood and the ark and it is proven to still exist among the children of Noah through their blessings particularly in the oldest son Shem as Japheth received this blessing that he would live within The tents of Shem."

A careful study of this text does not indicate a complete rejection of the model of the globe simply the concept of those who live in the opposite part of the globe. But, because of Saint Augustine's significance in the field of Church education and teaching, his words sparked animosity among some of the clergy towards the model of the globe.

Another important Church figure who believed in the notion that Earth can be flat, and floating in a vast ocean is Saint John Chrysostom (c. 349 - 407) Archbishop of Constantinople and one of the founding fathers in the Eastern Orthodox Church.

Cosmas Indicopleustes (6th century CE) was a wandering monk from Alexandria was the author of an important geography called

Christian Topography in 547, and in it, he described his Earth as a rectangle having an area that could take 400 days to traverse and the width of 200 days to travel across. The author wrote that there were oceans on each side that aren't accessible due to four massive walls that hold the sky. In spite of the fact that Cosmas traveled to India as well as Abyssinia, Cosmas based his geography on Scripture rather than his own observations. And despite being in the midst of the best of Western learning, he dismissed Classical arguments in favor of a round Earth as nothing more than pagan heresy. He fell victim to the trap that many flat earthers fall into discovering passages from the Bible that appeared to support the existence of a Flat Earth, and therefore any amount of scientific evidence could convince him that the Earth was not round.

A representation of Cosmas"worldview

Although these writers made a lasting impression however, the majority Christian scholars during the Middle Ages still believed the Earth was a globe in spite of the fact that Classical writings were found in limited quantity. A geography of Macrobius (early fifth century) depicted a spherical earth that was surrounded by planetary spheres and the work was copied several times in the Early Middle Ages in places like France as well as Ireland. Boethius (c. 480-524) used Macrobius to discuss what is the essence of Earth in his highly significant Consolation of Philosophy. The bishop Isidore from Seville (560-636) published in his bestselling publication, De Natura Rerum, that the Earth is a globe, with the Sun spinning around it, allowing daytime for one and evening on the other.

In his encyclopedia that is widely read Etymologies that was regarded by medieval scholars to be an extremely sophisticated works that ever was written, the Isidore stated that the sky was an elongated sphere with the Earth at its center. He also accurately described eclipses of the Sun as occurring when the Moon is able to pass between the Earth as well as the Sun. Lunar eclipses take place because the Moon is in an area of shadow that is dominated by the Earth which implies that the Earth is spherical. Earth.

Isidore's depiction of the zones that comprise Earth

The renowned English theologian Bede (c. 672-735) discovered support in Scripture for the existence of a globe-shaped Earth in using the Bible's references to "the orbs of the universe." A similarly significant Christian philosopher the saint St. Thomas Aquinas (1225-1274) taught it was true that Earth is an circle.

While this was happening there were attempts to accurately calculate the circumference of Earth. Hermann from Reichenau (1013-1054) was an influential German Benedictine monk who was later elevated to sainthood and discussed the measurement of Eratosthenes's the circumference of Earth according to the writings of Macrobius and acquiesced to the Greek's results.

Arab scholars, who preserved the writings of Classical writers better than Europeans in their own right, were acquainted with the early debates on how the world works. Earth and many Muslim academics through in the Middle Ages repeated that the Earth was an globe. For instance, ibn Hazm (994-1064) declared that, despite the common belief of believing that the Earth was flat every evidence pointed towards the Earth being a sphere, and that nobody should consider themselves an imam when they claimed contrary.

The long-running legend that has become a staple of popular cultural culture throughout the West is that Christopher Columbus sought to prove that the earth was round however, the idea that ignorant Spaniards and Portuguese in the fifteenth century believed that there was the existence of a Flat Earth seems to have been first introduced by American journalist Washington Irving, whose 1828 account of the explorer's life romanced Columbus by portraying his character as a visionary proponent of science that was empirical against the backdrop of the medieval obsolescence.

This description was at best excessively exaggerated and, at worst, an absolute lie, as Columbus certainly did draw certain evidence of the possibility of his ideas from the scientific findings of Portuguese explorations however, he also relied heavily from the older sources, such as The Bible, Aristotle, Ptolemy, Pliny and the more modern Cosmography from The Cardinal Pierre d'Ailly. Columbus had a real disagreement with his contemporaries involved another issue that was the actual diameter of Earth and the extent of its surface comprised of land and water. Different geographers and cosmicographers had reached different conclusions. There was some confusion over the estimates of Eratosthenes due to the differing measurements and systems that were used from Greeks, Romans, Arabs and even medieval Christians. The confusion led to Columbus's estimation of a smaller distance between westernmost tip of Europe and the eastern edge of Asia that was based on what he had read.

Columbus eventually concluded that Asia was at the distance of less than 2500 miles. This is what makes Irving's portrayal of Columbus as a scientist very absurd. In the end, the actual distance was 12,500 miles, making Asia at a distance of four times farther away than Columbus had anticipated. The historian Edmund Morgan noted, "Columbus was not a scholar. But he did study these books and made a myriad of notes in them, and came up with concepts regarding the universe that were simple and strong , but sometimes incorrect, the sort of thoughts that a self-educated individual gains from their own studying and sticks to, despite the things that others try to teach him."

Columbus Notes in Latin in the margins of the copy of The Travels of Marco Polo

The people who mocked Columbus are far from uninformed Flat Earthers were working with highly accurate measurements and had a good reasons to believe that Columbus's expedition would run out of food and water before it reached Asia. They lacked only one key fact that was the existence of a second landmass that was located between Europe between Europe and Asia.

In simple terms in other words, if Columbus had not discovered the Americas then he might not have been able to sail over the edge of the Earth however, the crew and he probably would have been forever adrift and never heard for the rest of their lives.

Chapter 2: Revision Of The Flat Earth Theory

In during the Middle Ages, the majority of theologians and scholars believed that the Earth is a globe as well as during the Renaissance as well as in the current there were a few people who had a reasonable education doubted whether the Earth was one big globe. Although Columbus might not have been accurate in his assessment of the globe's circumference, his voyage of 1492 was the one that was most influential in into the Age of Discovery, and when European powerhouses gathered all over the globe the globe model was well-known to general public, assuming that they believed in it.

At the turn of the century there were some eccentrics who began to suggest that the earth was flat, and others believed that even if the globe existed but it was hollow with plenty of space to live in. The rise in literacy along with lower postage and lower printing costs meant that there was a boom in various popular books and, in an era which embraced New Age beliefs, conspiracy theories, and various religions and cults publications and pamphlets about weird theories were able to find an public.

One of the early advocates of Flat Earth was Englishman Samuel Birley Rowbotham (1816-1884). A naturally intelligent and quick-witted person, he was a school unsuitable for him and quit at the age of nine. He was extremely religious and believed in the Bible seriously, and at an early age, he was fascinated by Owenism an socialist utopian movement that established many communes, including one which Rowbotham established.

Rowbotham

At this point he began to think of new ideas about the structure of Earth and then decided to test these ideas. The first test was conducted in the Bedford Level starting in 1838. The drainage canal, which has calm waters and a 6 mile (9.7 kilometers) stretch that is straight like an arrow was the perfect place to measure the curvature (or flatness) of the Earth. The concept was easy and was already in use that - the researcher would use an instrument to look through a telescope at markers that were laid out across the level. As explained by Rowbotham "If there is an earth that's a globe and has a diameter of at 25,000 English statute mile in diameter. The surface of any standing water must be a certain amount of convexity. Every part has to be an arc that forms a circle. From the highest point of such arc, there will be an angle or declination of 8 inches in the initial statute mile. For the mile after that,, the drop will measure 32 inches, and for the mile in third, it will be 72 inches or 6 feet. ..."

Rowbotham was in the river , with an instrument that was held at 8 inches (20 centimeters) above the surface of the water and watched a rowing boat go away from him, with an American flag on its mast three feet (91 centimeters) above the water. The boat was able to row six miles away from him until the bridge. Should the Earth was curving then the boat must be out of sight However, Rowbotham claimed to have been able to clearly see the boat throughout its entire journey.

The results convinced Rowbotham he had a point which is why in 1849 the results were published in an article (later expanded into the form of a book) called Earth Not A Globe. He went by the pseudonym Parallax as he was perhaps afraid of the repercussions of using his actual name.

As time passed, Rowbotham began to lecture across the nation about his theories about the Flat Earth, and at the beginning, things didn't take off as well due to the fact that the lecturer had not considered all of the possible arguments. One time, during a lecture in Blackburn one from the audience wanted to know to explain why that the ships' shells and hulls disappeared. vessels were unable to be seen over the horizon prior to masts. In a state of confusion, Rowbotham ran out of the lecture hall. However, he quickly improved showed a knack of thinking on his feet, and then burying the most rational objections beneath a sea of scientifically-sounding responses.

In 1864, a group of scientists decided to prove him wrong completely by asking him to go to Plymouth shoreline and view an telescope at Eddystone Lighthouse 14 miles (22.5 kilometers) from the sea. From the distance, because of its curvature Earth only the highest point of the lantern could be seen and the remainder of it was hidden beneath the sky. Rowbotham was then able to turn around and looked through the telescope and said it was possible to see all the lighthouse. After the scientists said there was no anything, he confidently claimed that he could, and said that it proved there was no evidence that Earth had a flat surface. He was so confident of himself that the majority of the audience left believing that he had won the argument.

The scientists were taught life lessons that people are still struggling to understand lies travel quicker than the truth and if repeated enough times and with sufficient conviction is able to alter people's thinking. Scientists can try to make arguments using evidence and logic however, if the other scientists operate with a totally different philosophy, they are impossible to reach.

Rowbotham began to collect supporters and, in 1870 one of them John Hampden, offered a wager that no one would challenge Rowbotham's achievements in his Bedford Level. A editor from a magazine for sports known as The Field, smelling good publicity, offered to serve as a referee. In the end, Alfred Russel Wallace, the naturalist and surveyor decided to take Hampden to court the bet. Wallace was aware that atmospheric refraction can cause light to bend near the Earth's surface and especially in regions that have temperature inversions, like those you see over water. Therefore, to minimize this effect, he placed the sightline at thirteen feet (4 meters) above the water's surface and placed an object in between the Bedford Level in order to determine the curvature that the Earth as observed from two different ends. Wallace's test proved it was true that the surface Earth did indeed curve , and the magazine gave him with the prize. However, Hampden demanded that the prize was returned insisting that Wallace was cheating and in all likelihood, he had to withdraw his offer before the test was even started. The court case was a long hearing that concluded with a ruling that Wallace was required to return the money as Hampden had actually withdrawn the offer. It was a cold comfort for Hampden who was in prison for threats to murder Wallace.

In the meantime, Rowbotham continued to give lectures on the theories he had developed, which became ever more complex as time passed. The professor referred to his work as "Zetetic Astronomy" and as the book explains, "The term Zetetic is an abbreviation of the Greek verb zeteo meaning to seek for, or to study and to conduct only investigation; not to take anything as given, and to investigate phenomena and their own immediate and obvious cause. It is here used in contradistinction from the word 'theoretic,' the meaning of which is, speculative--imaginary--not tangible,--scheming, but not proving." Rowbotham was a Biblical literalist, taking the Bible as one hundred percent factual and thus the mentions of the Earth being flat must reflect the truth. Below the oceans of the planet is a vast expanse of burning fire (that is why it doesn't cause heat to the water) which is the location where Hell is. Heaven is higher than the stars.

The theory of Dr. Rowbotham stated it was that Earth was flat, that had its North Pole at the center and all the other lands enclosed with the Antarctic Ocean bordered by a vast cliff of ice with a pristine land of ice and snow that stretched to the limit of. In reality, the Sun, Moon, planets and stars are hundreds of miles off that surface with the Sun traveling around at the North Pole once a day. The Moon is illuminated by its own light, and eclipses result from a mysterious, dark object that is spotted close to the different celestial objects.

This theory can cause a lot of issues. In the event that Earth is flat, with it's North Pole at the center then the continents on the south will have to be more distant than mariners had previously proven that they were. What is the reason why stars fall into Earth? What is the reason why is it that the Moon exhibit phases when it creates its own luminescence? Why can't its Sun visible from every point of the Earth in all times? What is its function as a light source?

Rowbotham attempted to explain these oddities using inadequate methods. In order to explain the visible Sun as he saw it as being moving across his North Pole, he asserted:

"The inquiry, 'how is it that earth isn't all times illuminated across its surface when the sun always rises hundreds of miles higher than it could be answered in the following manner"

"First If there were no atmosphere, then it is certain that the light from the sun would be reflected across all of the earth at once and the alternations of darkness and light would not be possible.

"Secondly because the earth is covered by an atmosphere that is many miles in its depth and the density of which slowly decreases until it reaches at the top, light rays, with the exception of those that are vertical when they reach the upper layer of air are held during their diffusion and then reflection, are bent downwards towards earth. Since it happens throughout the sun, and in all cases where the density and other conditions are the same and vice versa--the result is a distinct and distinct disc of sunlight."

For sunset and sunrise, Rowbotham offered the following explanation:

"Although that the sun appears to be always above the surface of the earth, it appears to rise from the north-east, to the noonday point before descending and disappear, or even set towards the northwest. This phenomenon is due to the simple and always visible rule of perspective. A large group of birds as they fly over plain or marshy area will always appear to lower when it recedes. when the flock is large it appears that the first bird is closer or lower towards the sky than another despite being in the same exact altitude over the earth directly below the ground. If a balloon is soaring away from the eye of an observer without altering or reducing its altitude it appears to slowly move closer to the surface. In a lengthy row of lamps, the third --assuming that the observer stands at the start of the row--will be lower in comparison to the previous while the third appears lower than the first and so on until the point of no return and the one that is farthest away will always appearing the lowest, though all of them have the same altitude. Moreover, If an uninterrupted line of lamps was extended long enough, the lights will eventually descend seemingly, to the horizon, or at an eye level of the observer ...

"Bearing to mind these previous phenomena , it is easily noticed that the sun, though always over and parallel to the earth's surface, can appear to rise out of the morn horizon until the noonday or meridian point and then descend until the evening's sunset."

Then , there's his explanation for the common phenomenon that mariners have been observing since early times The hull of a ship will disappear before the masts are. The subject had been a source of ridicule previously, but he has an answer. declaring:

"To say, for example that since the lower portion that is a vessel heading outwards vanishes before the mast head, the water has to be round is to believe that a surface that is round could produce this effect. If it is demonstrated that a basic law of perspective with a flat surface results in this kind of appearance, an assumption that rotundity isn't needed, and all of the misguided fallacies and confusion in or entangled with it could be avoided.

"Any distinct portion of a body will be invisible until the whole or any greater part of the same body is proved. Therefore, it is easily realized that the hull receding ships that obey the same law has to disappear on a flat surface prior to that mast's head. If it is placed into the form of Syllogism, the result is obvious that is:

"The hull is an important component of a ship.

"Ergo the shell of a receding or outward-bound ship will disappear in front of the entire includes the mast head.

"To provide the argument with an even more practical and nautical slant, it could be stated in the following manner:

"That portion of any receding body closest to the surface that it is moved, contracted and is in-visible to the parts further away from that surfaces ...

"The the hull of a vessel is closer to the water, or the surface that it is moving on, than the mast head.

"Ergo the ship's hull that is bound for the outward ship is the first vessel to vanish."

This assertion even though it appears to be seeming sensible and rational however, it is clearly false. If any thought is put into it, the conclusion is utterly absurd.

The remainder of the book continues in a similar manner even though it is a poor argument, his theories had a huge impact in pseudoscientific and alternative religious circles. Perhaps, he can be considered to be the creator of the current Flat Earth trend. Its Zetetic Society even opened a chapter in New York, and Rowbotham delivered a thousand copies the book to New York in order to give to Americans.

The work of Rowbotham led to the acceptance the Flat Earth beliefs by the Christian Catholic Apostolic Church. Established in the year 1896, by John Alexander Dowie (1847-1907) who was a Scottish spiritual healer. the church rejected modern medicine. It also banned alcohol pork, short skirts or whistling at the church on Sundays. The church was successful and Dowie later founded Zion, a town in Zion located in Illinois to serve his faithful. Zion rapidly expanded to 6000 and Dowie began to wear elaborate gowns while using the name of "Elijah the Restorer." Many of Zion worked for Zion Industries producing items ranging from fruit-based bar to Scottish lace and Dowie himself splurged on the funds. In reality, he was spending so many times that the church was into bankruptcy. In 1906, his faithful ejected Dowie in favor of Wilbur Glenn Viva (1870-1942).

Dowie

Voliva

Voliva was the person who brought the concept of the Flat Earth into the church's doctrine. Church schools insisted there was no evidence that Earth was flat. The authorities went so far as to say that globes were not allowed within city boundaries.

Despite his strange opinions, Voliva was a shrewd businessman and was able to get the congregation back to solid financial basis. The church in 1923 launched the radio station WCBD which was the name of the station. Voliva was the first preacher to have an own radio station. In the next few years the entirety parts of North America and even parts of Central America were treated to lectures on the dangers of astronomy and evolution and all things science.

Voliva had been a Bible literalist who mixed his narrow interpretations with moderate amount of humour. One time, he stated, "The idea of a sun with millions of miles the diameter, and nearly 91,000,000 miles distant is absurd. The sun is just 32 miles wide and is not greater than 3000 miles away from the surface of the earth. This is why it's so. God created the sun to illuminate the earth and should have put it close to what it was made to perform. What do you think of the man who built a home in Zion and then put the light bulb to illuminate the house in Kenosha, Wisconsin?"

His theories were a normal Flat Earth model, with some added religious nuances. As per Voliva, Earth was a disc that was located at The North Pole and ringed with the ice barrier has been mistaken as Antarctica. The ice-free zone was Hell and under it was a place that was inhabited by ghosts of a race who existed on Earth prior to Adam or Eve. Sky was like a dome that had stars that were fixed within it. The Sun as well as the Moon were both bright and much smaller than Astronomers (whom Voliva called "poor, ignorant, arrogant fools") declared to be.

Each evening, Voliva would sit before the microphone at his radio station at home and protest against the establishment of science, saying Voliva could "whip into smithereens anyone anywhere in the world in a mental struggle I've not met a single professor who knew one millionth of the information on any matter like I did." People across the globe were equally amazed and amused.

As many preachers before and in the past, Voliva also predicted the end of the world. He began by predicting that the end of the world in 1923. When it didn't happen then he changed his date back to 1927. He then moved it up until followed by 1930, 1934 and finally, 1935. In his own words the future, he vowed to remain alive until the end of time and in actuality, he declared he would get to the age of 120 because of his strict diet, which predominantly, consisted of buttermilk as well as Brazil nuts. He fell 50 years behind, passing away in 1942, at 72.

As with many preachers, including his mentor, Voliva began to amass an enormous personal fortune, at the cost of his church, and both the church and he went in financial ruin. Amazingly, the church survived and is still in existence in the present, however it has stopped promoting it's Flat Earth theory and is today more of a traditional fundamentalist church.

It is possible to believe that when Rowbotham passed away in 1884, his own self-made discipline that was Zetetic Astronomy would die with the deceased, but he was surrounded by faithful followers who carried on his principles. Most important lady was Elizabeth Ann Blount who was the founder of The Universal Zetetic Society with the goal of disseminating "knowledge about Natural Cosmogony in confirmation of the Holy Scriptures, based on the practical investigation of science." That is to say the purpose of scientific research was to verify what the scientists had already concluded was the truth and was in complete contradiction to the way science actually works. The society's journal was named The Earth Not a Globe Review and ran for a number of years. Lady Blount also published Earth: A Monthly Magazine of Sense and Science between 1901 until the year 04.

Lady Blount

The Earth Not a Globe Review included a shrewd mix of polemic, lengthy articles, letters to the editor and critiques of the science reporting of the mainstream press as well as scientific education at schools. The May 1896 issue, titled "Faith as well as Science" declared that "we are determined to expose the complete falsity and unscientific nature of the modern theories of geology, astronomy and evolution. We will also show that they are all one and all, and not just non-scriptural, but also absurd and non-philosophical. We call on the most competent scientists of our time to defend their preposterous theories, and the theories based on them or to discover any flaws that is not in their Divine Cosmogony of Holy Writ."

The article questioned the validity of a Christian magazine, The Faith, in which an article attempted in resolving the issue of evolution with Biblical account of creation affirming that creation was simply a "series of events." The Earth Not a Globe Review opposed the idea, "We reply, not necessarily. The "divine purpose" mentioned in the Scriptures is not a 'series of processes', however it explicitly and firmly teaches the creation to have occurred instantly. The line that separates error from truth is a exquisite one. And this claim of a'series processes" is the very first step in Scripture. Scriptures of Truth on the path to falsehood and error."

The article continues to claim the absence of evidence of evolution either in the past or the present as well as that, since Adam became the first human being and Adam was the first man to be born, there is no room in Biblical theology for an "pre-Adamic person." The article concludes with a song:

"There existed an ape back in the earlier days;

As time went by, his hair curlier.

More than a century later, he added a thumb to his wrist.

Then, he was a man and one who was an Evolutionist."

Another major fan who was a major supporter of Rowbotham is William Carpenter (1830-1896), an English printer and stenographer. Carpenter was instrumental in helping Rowbotham to create many of his pamphlets and books and, in 1879 the year he left England for Baltimore to become one of the most prominent Rowbotham supporters throughout his native United States. He also became a vocal advocate for many other causes considered odd in the past like eating vegetarianism, hypnotism, or spiritualism. With the pseudonym "Common Sense" he published a variety of pamphlets about these and other topics, such as two books about flat earths. Flat Earth: Theoretical Astronomy Examined and Exposed Proving that the Earth is not is a Globe (1864) as well as The Hundred Proofs that the Earth isn't an actual Globe (1885).

The proofs are a hundred and include the following:

"8. If the Earth was a globe, it is probably the best--and the only thing a navigator would need to bring out to sea. But there's a truth that is not understood that with an instrument as a reference the mariner would destroy his vessel, with certain fact!

This is proof that Earth isn't an actual globe."

"15. The notion that in lieu of traveling horizontally around the Earth ships are taken to one side of the globe, and then under and re-emerged on the other side , to make it back home, is, aside from a mere fantasy ridiculous and unattainable! In the event that there aren't any impossibilities, nor absurdities in the basic idea of circumnavigation, it is as a fact that the Earth isn't a globe."

"17. Human creatures require a surface on which to live, and that is, in its broad sense is LEVEL and as the Omniscient Creator is completely aware of the needs of His creatures and that is why the All-knowing Creator He has met their needs completely. This is a theological evidence to show that Earth is not an earth."

"62. It is generally believed that the Earth should be a globe since people have been sailing around it. Since this means that we could travel around anything unless it is a globe, and it is acknowledged that we are able to travel around the Earth in a plane, this assertion is absurd and we have yet another evidence to prove that Earth isn't an actual globe."

"72. Astronomers inform us that due to the"rotundity" of the Earth, the walls of buildings that are perpendicular arenot, in any way, parallel. Even those walls that are that are on the opposite side of streets not! Butsince every observation does not provide any evidence for this deficiency of parallelism that theory claims that the theory be disregarded as absurd and contradictory to all the well-known facts. This is evidence of the fact that Earth is not an actual globe."

Chapter 3: "The Flat Earth Movement In The 20th Century

The concept of the Flat Earth endured throughout the 20th century. And even while it was at the margins, scientists continued to create societies and publications on the subject. They were more likely to repeat theories of the earlier writers like the assertion (to which Ptolemy was acquiescent) to the effect that Earth cannot be spinning as the Earth would experience a terribly strong breeze all the time.

Gabrielle Henriet, in her 1957 book Heaven and Earth in her book Heaven and Earth, asserted that the speed of rotation is calculated by astronomers as the rate of 620 miles/hour (1,000 km/hour) which is about how fast an airplane in the day. She claimed that "an aircraft that is flying at this speed at the same rate that of the rotation would never cover any ground even if it tried. It would be suspended in mid-air, over the location where it began its flight as both speeds are the same. Additionally, there would not be any need to travel from one location to another in identical latitudes. The plane could start to rise, wait for the country you want to visit to appear in the normal course of the rotation and then take off; however, it's difficult to imagine how any plane would be able to even touch the ground at an airfield that is moving at the speed of 1,000 km per hour. It could be beneficial to learn what the people flying around think about the motion that is taking place on Earth. Earth."

The reasoning is easy. The atmosphere spins along with the Earth as the planet is moving through an empty space that is not a friction. Moreover, the Earth's gravitational field ensures that the atmospheric field is moving along with it. A good analogy to illustrate this is sitting in an enclosed car that travels at 100 miles an hour. If you throw a ball and it doesn't hit into their face since the ball is moving in the same direction that the driver.

Henriet was also adamant that the mainstream science was incorrect in the explanation of the seasonal change. It is not because of the tilt of Earth's axis because the taller buildings move between the sides.

Following the dissolution of Lady Blount's Universal Zetetic Society, Flat Earthers with the exception of the minor religious sects did not have a formal organization to function, however this was solved with the help of Samuel Shenton (1903-1971), an English sign painter who established in 1956 in 1956 the International Flat Earth Research Society. He was influenced to create The Zetetic Society, and his conception of the Earth was similar to the Zetetic Society's model. According to his theory the universe was vast and Earth being the Earth as a pit on an otherwise uninteresting plane. The Sun was like the flashlight of the sun, shining light to specific parts of the Earth during the 365 days of the year. This Sun has a diameter of approximately 32 miles (51 kilometers) in diameter and 3000 miles (4,800 kilometers) over the Earth. The Moon is approximately the similar size to the Sun but is closer, to 2,550 miles (4,100 kilometers) above Earth.

Keith Poole's photograph shows Samuel Shenton giving a lecture

Similar to the Flat Earthers prior to his time, Shenton came at the idea from a religious standpoint in a way, refusing to accept the idea of a round Earth, when the Bible clearly portrayed flat earths. To create his new society He did tone down the religious language to ensure it had the widest possible audience, and he was also less adamant than most Flat Earth leaders. He didn't make himself president of the new society, but instead named himself secretary and then naming the aging William Mills, who had been within Lady Blount's circle to be the show's director.

Shenton was a constant proponent and advocate of his theory, speaking to any audience, no matter how small , and frequently appearing on radio and television. Shenton was a lively and likable speaker who gained admiration from numerous within the scientific community and even the astronomer Patrick Moore, who ran the popular TV show The Sky at Night. Moore and Shenton were often at odds and Moore truly enjoyed Shenton's company, even when He slammed Shenton's theories.

In the second part into the 20th century Flat Earthers were faced with a new adversary that was the astronauts. The Soviets created the initial satellite Sputnik on the 27th of May in 1957 however, it didn't stay the only satellite for very long. within a short time afterward, people in the Flat Earth movement had to explain the images of the globe that were returned by spacecraft. At the end of the 1960s, the people required be able to describe the Moon landing and the way in which the weather satellites and communication were working.

The initial reaction in response to Sputnik in The Flat Earth Society in England was that a satellite orbiting over the Earth didn't negate their hypothesis. The Moon moves above Earth with a much higher elevation. Samuel Shenton quipped, "Would traveling around on the Isle of Wight prove that it was round?"

The sentiment wouldn't last for very long, since a couple of years after, astronauts were orbiting the globe, and later going towards the Moon. Communications satellites circled the Earth and weather satellites also sent back thousands of photos of the world. The most frequent reaction to these issues is that they were fake as well as that Soviets as well as the Americans and later Europeans, Chinese, Indians and other nations were making up myths about sending satellites and humans into space. This was part of an elaborate scheme to deceive people into believing that the Earth was flat. the Earth.

The reason why governments around the world would be willing to do this is somewhat of a mystery. Many Flat Earthers believe that they don't know what they're hiding from the authorities. Some say it's part of a mind control program or that space agencies had to justify their excessive budgets. Some claim that since governments have been caught in the lies they aren't able to acknowledge the truth and not lose their credibility.

Flat Earthers also have to understand why the thousands of employees who have been employed by NASA or other space organizations throughout the years have ever revealed the truth. One might think that a disclosure like this could lead to instant fame as well as lucrative book and TV deals. It is generally believed that a mixture of fear and hefty pay packets keeps employees on the right track, but others defend this by saying that there aren't a lot of people on the inside scoop. A commentator in the Flat Earth YouTube video put it, the majority of NASA workers are either janitors and security guards. How many of them are willing to lie?

In a strange way, at the peak of the Space Race in the 1960s The Flat Earth Society saw a increase in membership due to the attention that came from the latest astronomical achievements and from Shenton's appearances on the media. Following the Moon missions of The Soviet Union and the United States However membership fell off and it became more difficult to support this Flat Earth model in the of overwhelming evidence. The year was 1966. Shenton released The Plane Truth A pamphlet with a stated goal was to "re-establish among young people faith in God and a more accurate understanding of their earth's ecosystem." In the pamphlet the pamphlet, he wrote "Many would like to know more about the Society. It is true that today, very few people have access to it in number. As you could imagine, being against accepted Godless thinking and the devilish method of indoctrination, which instills theories that are not proven to be true in the heads of young pupil or student and no money, grants or aid is accessible to us. A couple of very noble men, now aging are still members of the previous Universal Zetetic Society." Shenton encouraged his readers to research the Moon missions from the perspective of Zetetic philosophy to determine the real meaning behind what could be a major misinterpretation of the ignorant scientists.

A letter addressed to the editor of The Plane Truth reflected this worry about godless thinking:

"The name that was given to the 'Moonship', by someone from the N.A.S.A. organization was 'Apollo'. It was one of the greatest divine beings from the Greeks. Every sudden death was thought to be the result of the Arrows of Apollo. It was reported on 27.1.67 The sudden deaths of three American astronauts while testing the space vehicle at the firing place located at Cape Kennedy. Another time, Florence, Italy, was suddenly struck by the torrent of ARNO overflowing. The Apollino symbol of god is linked to Florence.

"The idea behind these remarks is that more care needs to be taken when putting names to these automobiles ... We cannot belittle God in any way we want. I propose that the long-standing insult to God by claiming that we would rather believe Godless men about the existence of the earth that we live in and 'placed', has yet to be dealt with by us!"

Shenton when he was discussing the first Apollo flight (before that Apollo 11 landing in 1969) He repeated the error of many Flat Earthers prior to his: "I would like to bring our attention towards the facts of the 5.500

miles within the Atlantic test area and the time of 39 1/2 minutes to finish the splashdown. The Apollo craft had to travel across the Equator. The way we teach this is turning at a rate in excess of 1.000 m.p.h. A precise shot to reach the target that is taking away from the the initial estimate of location. There is no need to discuss or exaggerate the thoughts that are triggered by any reference to the supposed orbital speed that is believed to be theoretically possible. It will be evident that the Polar projection map shows that the preprogrammed orbits of capsules or missiles don't go over"the ice border" which we've not made progress. The moon, sun and other celestial bodies are within the same spheres close to the earth's surface however, they do so in an east-west motion. Most satellites made by humans are launched in the direction of west-east."

The pamphlet also provided an illustration that showed the Flat Earth with the "orbits" of the missions manned by humans shown as elliptical lines over the map, which explains what the pictures were captured. It is interesting to note that Shenton doesn't claim that the missions were fake, which is a conspiracy theory that will be discussed later on.

Shenton also provided an explanation of why the curvature of Earth can be clearly seen in a number of photographs: "Remember that all reproductions of images taken from capsules or rockets, are the way the camera recorded the event, and not according to the astronauts' humans. The camera's distortion of horizons has always been a source of confusion for those who haven't let go of the 'Globe' or 'Planet Earth' training. A few years ago when the U.S.I.S. publication 'Science Horizons ... contained an explanation that the Americans were hoping to develop an optical system that would not alter the horizons of level. As of yet, I'm unaware that this assistance to photography that is truer has been released. Flat Earthists can however prove that due to the established principles in perspective the horizon optically, increases and remains in line with the viewer's or camera's regardless of what height is attained. In reality, the earth immediately below the balloon, plane or rocket, appears dish-shaped or has a concave look. The area of earth directly beneath the vehicle is the lowest point. It's not the highest point on your "globe" earth. the curve or dip of the "ball" moving down to a unimaginably distant horizon below eyes!"

Her death in 1971 snuffed out the most important asset of the organization. At the time the number of members of the International Flat Earth Research Society had dropped to less than 100. In fact, Shenton predicted before his passing that the group was likely to die due to relentless attacks from mainstream science.

At first, this seemed to be the situation. Ellis Hillman (1928-1996) took the reins, but did very little to improve the society as his goal was to make a name for himself within Labour Party politics. The group therefore went on to fail.

Luckily for Flat Earthers assistance was provided by the name of Charles K. Johnson, an aircraft mechanic in California. The widow of Shenton was not sure about Hillman who she viewed as an opportunist.

She offered Johnson the majority of her husband's library. Johnson established The International Flat Earth Research Society of America and, as an avid Biblical traditionalist, he established The Covenant People's Church, a fundamentalist sect.

Johnson was more energetic and had more resources than the poor chronically ill Shenton who was under his guidance, the movement continued to expand. He published a constant stream of publications and a weekly journal called Flat Earth News, and Johnston was the first major public figure to declare that on the fact that space was just a fraud. According Johnston, according to Johnston that the governments around the world were trying to divert masses away from truths of the Bible with a huge campaign of false news. He also claimed they claimed that American as well as the Soviet governments were working together with this fraud to gain a profit for them and the elite companions. Johnston claimed that the Moon landing was recorded on a set and written by the science fiction writer Arthur C. Clarke. Johnson also was vocal about evolution and other findings from science that do not contradict Scripture.

Flat Earth News Flat Earth News was a combination of fundamentalist polemics as well as attempts at reasoning with science. The group developed an entirely different version that depicts The Flat Earth, with the North Pole at the center and the surrounding continents and surrounded by an ice wall 150 feet (45 meters) high. It is estimated that the Sun as well as the Moon are both 33 miles (52 kilometers) in circumference The Sun at 3,000 miles far from its surface. Heaven is a little further and the Moon just a little closer.

With dedication and a flair of causing controversy Johnson created the movement into an extent that was greater than it had ever been before. He claimed that the membership of the society reached 3,500. Johnson was a fan of publicity, and described those who believed in globes as "Greaseballs" due to the fact that they would slide off the Earth in the event of the shape of a globe. He was fond of showing photos of his spouse in Australia in which he pointed to the fact she was standing upright and not upside-down like they would be in the event that Earth was the shape of a globe.

The tragedy struck in 1997, when a fire in the house caused the destruction of his membership list as well as some of his publications. His wife passed away the same year and Johnson went to her tomb in 2001.

Flat Earthers didn't have to face hostile scientists and public ridicule and a hostile enemy from within. In the year 1970 there was a new society called the Flat Earth Society of Canada was established by its founding members, who included a professor of philosophy at St. Thomas University named Leo Ferrari, writer Raymond Fraser as well as poet Alden Nowlan. The Flat Earthers initially welcomed their counterparts from who were north as travellers however, the new society was determined to be more outrageous than normal Flat Earthers. They claimed that all of society's evils, from racism to misogyny, were because people followed the "globulist" model rather than being "planoterrestrialists" like them. Ferrari published a scholarly paper, "Feminism and Education in the Flat Earth View," that was published in McGill Journal of Education. The motto of the society was "The Earth is flat; any fool would know that."

Ferrari and Nowlan appeared in several appearances on the media and soon began to draw attention. However, since the majority of the members originated mostly from United States, they changed the name of the society and changed it to The Flat Earth Society in 1973. The following year the society was home to around 100 members.

The activity continued to be strong through the 1980s, however many did not believe that Ferrari and his associates were sincere. They injected a bit of humor in everything they did and that was something they did not have in the more sincere Flat Earthers. When the group was founded in 1990, it was involved in helping create film In Search of the Edge produced by Pancake Productions, and red warnings were raised in those in the Flat Earth community when it was discovered it was it was the Ontario Arts Council and National Film Board of Canada were aiding in the financing. In the end, how could an unfriendly government who claims to be the one who launched satellites financing the Flat Earth program?

The suspicions were right. The film was a prank and was intended to be shown in schools to show children how absurd ideas can be made to seem more appealing when presented with conviction. The study guide distributed with the film explained it was a prank and its founding members are "globularists," and that everybody should develop their critical thinking abilities. The catalog description of In Search of the Edge stated:

"The film is composed of the standard documentary-style mix of elements like interviews with experts and stock footage, as well as stills and animation and is interspersed with the tale of Andrea Barns, a Canadian woman who has dedicated her entire life to convincing the world that the earth is not flat.

"This is a crucial instruction in the art of media literacy to the generation that is connected to television. Inspire critical thinking in your students and guide them become aware of the tools employed in documentary media as well as the development of scientific theories as well as their own determination to not take things at their own worth."

Naturally those who were Flat Earth advocates were not surprised. The Flat Earth advocates were among the people who consistently asserted that other people took the facts on faith and that the claim to be made against them was unsettling.

Chapter 4: The Internet

The Flat Earthers continued to be at the margins of society throughout the 20th century, however a brand new technology gave their cause an opportunity to gain momentum with instant global communication on the Internet. In a matter of minutes, those who believed that they were all alone in their beliefs were able to form groups to discuss their beliefs and gain an ear to listen. Discussion groups discussing conspiracy theories arose along with"the "information superhighway" quickly became the most powerful tool to spread ignorance.

The Flat Earth Society, which was in decline after Johnson's death in 2001 it has been revived by Daniel Shenton (no relation to Samuel Shenton), and the society was given its new shape, based into the Internet discussions forum. The forum increased to thousand members. It also launched the wiki and also included sections that covered a wide range of Flat Earth literature both modern as well as historical. Since 2009 the Flat Earth Society launched in a new and official manner. In its press release Shenton declared that although traditionally the society disseminated its message through pamphlets and newsletters, "Technology has moved on and we're happy to be capable of reaching a larger public by delivering our message. Internet is a fantastic tool for communication. Internet is an incredible tool to reach out ... I'm convinced that we're getting the attention of the public and that's extremely important. I realize that our beliefs aren't always accepted as orthodox But people are starting to inquire and that's the first step in the process of seeking truth."

Shenton was completely right on the impact instant online communication could be able to have on The Flat Earth movement. In actuality in the very first instance ever, it actually became an entire movement. Many people took part in online forums and it didn't take long before the podcasts and YouTube channels became part of the equation. People who not heard of these theories were exposed to them. The media coverage helped ignite the flame and a growing amount of celebrities began to support theories like the Flat Earth theory. Every time the same, there would be a media frenzied and people would look up this Flat Earth movement out of curiosity.

There are numerous websites and forums online that promote that Flat Earth view. The most well-known is The Flat Earth Society, which continues to thrive with a forum that has hundreds of thousands of posts as well as an extensive website that answers the most frequently asked questions regarding the theory, the most famous people as well as documents from founder fathers like Rowbotham Carpenter and Rowbotham. Carpenter.

Twitter also has a lot of lively conversations that are lively, with Flat Earthers frequently mocking people who think that they believe that the Earth is round by using the derogatory hashtag #globetards. A lot of the content on Twitter is self-reinforcing, and users are exchanging jokes about how ignorant their fellow members are. This is a typical feature in Flat Earth online communities, in which members spend the majority of their time applauding each fellow members for seeing through society's illusions and revealing the reality.

However, they're not the only ones to be laughing. A lot of people online make fun of Flat Earthers. There are a variety of fake Twitter feeds, such as those from the Dinosaur Earth Society, which asserts that the Earth has the form of an animal and challenge anyone to disprove this claim.

One of the most effective methods of attracting people to people who are part of the Flat Earth movement is YouTube which has seen an explosion of podcasts as well as amateur documentaries that promote on the Flat Earth idea. A majority of these have amassed up thousands of views and the most well-known channel is owned by Mark Sargent, who has more than 77,000 subscribers and over 16 million views total. The popular YouTuber frequently uploads interviews, podcasts and reports on meetings for the numerous local and international Flat Earth conventions that happen throughout the world. Sargent has become a household name within Flat Earth circles. Flat Earth campaign.

Another popular figure Another is Robbie Davidson. While Sargent is a proponent of global theory, mixing conspiracy theories and free thinking, Davidson is firmly in the fundamentalist Christian camp. The description of the YouTube channel he runs, Celebrate Truth (which has more than 136,000 subscribers, and more than 19 million page views) says ""To spread Christ's Truth as well as expose lies of the world. What if all that NASA & Science has taught you about the earth sun, moon space, and stars was an untruth? What if Satan trying to do to undermine the fundamentals of the Word of God by deceiving the truth about the flat earth that which we reside on since the beginning? This may sound absurd, but it's worth the effort to investigate this subject since it may surprise you with the biggest fact you've ever learned about the world we reside in. Go to your bible and read the entire chapter on God's creation. Evolution is a lie , is nothing more than a lie that claims it was an accident during a massive explosion. Monkey Man Science is all part of the big deceit. Don't be fooled by the fakes."

Flat Earthers are forming an online community, too. There are many Flat Earth Meet-up groups throughout the globe Conventions bring in hundreds, sometimes many thousands of people. There's even an official Flat Earth cruise scheduled for 2020. As the Guardian hilariously stated that marine navigation is based on the premise that the Earth is round and that's why GPS (the Global Positioning System) is accomplished by using 24 satellites orbiting around the Earth and ships require 3 GPS Satellites for triangulating its location.

Current beliefs

Fundamentalist Christians have been the core for the Flat Earth movement since its inception The increase in the amount of Christians who take the Bible in a literal way, and thus reject theories of science, such as evolution, has provided an ideal recruiting ground to members of the Flat Earth movement.

Flat Earth theory has gained some traction. Flat Earth theory has also gained some traction in The Islamic world. Similar to

Christians and Christians, they cite Bible passages that point to the existence of a Flat Earth as well as also refer to texts that are from their sacred texts and academics. Certain such texts have been included in a tiny anonymous pamphlet entitled Earth is Flat and There is no Dar al-Islam Today. It contains a number of verses in the Koran which talk about Allah spreading the Earth as "a beds" and making"the "canopy" over the heavens. The conclusion, of course is that the Earth is flat, and it is like a dome but there is no place in the Koran does it explicitly state that it is the case.

The pamphlet continues to include a variety of quotes from Muslim experts who believe that the universe is flat. For instance the pamphlet says, "Al-Qahtani said in his Nuniyya The astronomer as well as the soothsayer and all their followers are lying, they claim to have the knowledge of Allah. Both of them believe that the Earth is spherical, and they share the same statement. Also, the earth is flat to the intelligent people that are based on the authentic proof of the clar Quran." It is too humorously, this anonymous writer did not mention any of the respected Muslim scholars who had argued for a spherical earth.

The pamphlet also states that the Moon emits light, and doesn't transmit light back to the Sun. This is based on truth that Arabic word to describe the Moon is only meaning "light," not "reflection of light."

In a similar way to the notion about an Flat Earth being surrounded by mountains in a ring, there is talk of a mountain known as Qaf which surrounds the earth and is the point on where the sky's edge rests.

As a final proof as proof, the reverse of the cover depicts an unflat Earth as well as the globe Earth and Muslims praying to Kaaba in Mecca as fervent Muslims perform five times per day. On the globe Earth however, the Muslims don't face directly towards Mecca because of the curvature of Earth. Only on level Earth are Muslims capable of facing directly at Mecca in the way that Allah would have.

There are numerous videos on the internet of Islamic preachers who claim that the earth is flat. One Saudi preacher on an online video reiterated Gabrielle Henriet's assertion that the Earth cannot be spinning as when a person throws a ball in the sky, it travels straight upwards and downwards.

However most Muslim public figures believe that the Earth is an earth. A popular video Muslim journalist Sheikh Assim al-Hakeem reveals that the Koran does not speak of the earth as flat, and of the skies as a canopy because that's the way people perceive it. That is the Koran is not intended to explain how the world actually formed. The author goes on to explain that there are evidences from the Koran to demonstrate that the world is circular, and that scholars agree in their belief that this is true.

Another commentator who is the Dr. Zakir Naik, says it is now proven that there is no evidence that Earth can be described as a globe. He also discuss scientific errors in various chapters of the Bible previously discussed that, like Matthew 4:18, Luke 4:5, Daniel 4:10-11, First Book of Chronicles 16:30, Psalms 93:1, and John 10:30. These verses indicate that it is possible that the

Earth has a flat surface and is indestructible in addition, it is believed that Dr. Naik pointed them out to indicate that the Koran is the most reliable document. Koran is the most authentic and more reliable source.

It is not clear what proportion in the global Muslim population believes they're standing on the Flat Earth and, given the fact that Muslim media personalities are often confronted about the form of the Earth the concept seems to be getting more attention across the Muslim world, as well as in the Christian world.

Although it might seem, there are organizations comprised of Flat Earth researchers looking at proving their ideas with scientific research. Similar to when the Bedford Level experiments were used to "prove" that the Earth is flat a similar test was conducted recently by Flat Earthers as part of documentary Behind the Curve. The small team of scientists known as Globe Busters set up two boards on a large expanse of calm, shallow water. They made holes that were 17 foot (5.18 meters) over what is below the surface. One of the members stood in front of each of these holes and another stood a significant way away, using an illuminated flashlight. In the event that it were the case that Earth is flat it would not matter how the light was located from the boards as it could be able to reflect through the holes however, if the Earth is curved in the extent that scientists have claimed and the man holding the light had to keep it at a distance of the distance of 23 inches (7 meters) in order for it to shine through holes. Much to the Flat Earthers their dismay after the light was switched on and held at a height of 17 feet, it wasn't evident through the holes. The observer then requested the person holding that light up to the height of 23 inches (7 metres). The light reflected and confirmed that, due to how curvature is formed by the Earth the person who had the light was further down than the holes.

Another of the experiments featured in the documentary concerned the use of a Gyroscope. Gyroscopes that spin, because of the unique physical physics that they employ they will always be in the position from the beginning of their spin, regardless of the surface they're spinning on. This means that they "tilt" relative with the surfaces of Earth as the surface itself is being tilted.

Flat Earth YouTuber Bob Knodel spent $20,000 on an precise and well-made laser gyroscope that was used to check if there was evidence that Earth spins. With a spinning globe the gyroscope can move 15 degrees per hour due to the Earth's spinning and, when Knodel along with his associates conducted the test, this is exactly what they observed.

In both instances, the Flat Earthers didn't take the findings of their own research. It's likely that they would have jumped at the results had the results were in line with their expectations.

In the year 2018, one brave Flat Earther made the leap to a whole new degree. "Mad" Mike Hughes completed work on a rocket made from scratch and shot himself into the sky hoping to observe what lies beneath the Flat Earth for himself. The launch received a lot of attention from the media, including Hughes launch from a remote location within the desert of Mojave. The rocket, constructed from scrap metal that cost about $20,000, only soared 1,875 feet high before it deployed its parachutes before landing on the floor of desert. Luckily, Hughes emerged unscathed aside from a back injury.

Unfazed, he promised to build the best rocket possible, and the next time that he'd prove that it was true that Earth was flat for good once and forever.

Chapter 5: How Do Flat Earthers Think?

To take this off the table It's not an omen. There are many who sincerely believe that the Earth can be flat. One of the very first words from the Flat Earth Society website is "This website is not a parody. We are actively working to promote The Flat Earth Movement worldwide." It's not the best way to build confidence when your site has to make it clear that it's not a joke.

In any case, here's the list of their fundamental principles that they hold to be

1. The earth is flat.

2. Space doesn't exist.

3. The Earth is protected by an impassable Dome.

4. It is home to the Sun, Moon, planets and stars are located inside the Dome.

5. The Earth is enclosed by an Ice wall.

6. Gravity doesn't exist.

7. The Moon is self-illuminating.

8. The Earth is just about a couple of thousands of years old.

Let's take a look at these basic ideas one-by one.

1. The earth is flat.

It's not, but I'll look into it.

2. Space doesn't exist.

Based on the Flat Earthers the universe we have as isn't real. While the existence of black holes, nebulae supernovae, neutron stars asteroids, pulsars and galaxies were observed by numerous experts over the decades, conspiracy theorists insist this is all a flimsy lie.

This claim is absurd. Astronauts upload videos of their time at space stations on the internet each day. If space travel were fake the cost would be an enormous amount to construct structures, wires, upload computer-generated images and more.

If NASA had been faking it all then why are they streaming videos with astronauts floating around in space or in space shuttles? What's the reason to risk being found out? If NASA declared they were private, which means that the public wouldn't be able to be able to view them, the world would know. NASA is not under any obligation to reveal the actions of their employees. If the videos were faked the only thing it takes is one light off or a single mis-timed special-effect or just a single wire snap for an illusion of realism to be dismantled.

In addition, astronauts undergo numerous bodily changes as they spend a long time in space. A typical astronaut is affected by an impaired heart as well as muscle atrophy and the loss of bone, mostly in the neck and the spine. The long-term exposure to space may make an astronaut's heart spherical shape. Twelve astronauts who were examined prior to and following their space explorations, the ultrasound tests confirmed that their hearts were more spherical. 9.4 percent.

After a few months of space, the muscle mass may drop by 20 percent. Astronauts may lose as much as 1.5 percent of their bone density each month. Additionally they often lose their fingernails when they do an exercise in space. When astronauts have spent a full year in space, they're approximately two inches taller due to the fact that their cartilage discs are expanding due to microgravity. Flat Earthers cannot describe any other scenario where the body of a person undergoes such drastic physical modifications.

3. The Earth is protected by an inaccessible Dome.

The Dome is a reference to the "firmament" which is mentioned within Genesis within The Bible. Other religions also have the same concept. Naturally, the Dome is intended to be impenetrable in order to justify the belief that no one is able to penetrate the space. If the Dome cannot be penetrated, from where do meteors originate?

Despite Flat Earthers' belief that the Dome is a real structure, they aren't able to agree on exactly what Dome is comprised of. Metallic crystal. Dark Energy. Hydrogen. An unquantifiable element. It is unlikely that even the Flat Earth Society can give an answer definitively.

It's a common belief that Halley's Comet appears in the sky each 776 years. It first appeared in 1986, and is expected to return in 2061. Because the Flat Earthers do not believe that space exists, they believe Halley's Comet exists right beneath the Dome. Where is it? Why can't we tell it as distinct from the time that Edward Halley predicted it? Are they hidden somewhere?

Additionally, if the Dome is inaccessible, that means nothing can penetrate it, isn't it? There are numerous videos on the internet of weather balloons released into the sky. They have cameras attached to them that observes the absence of a Dome (and you may also observe a circular Earth as an added bonus.) This is the conclusion you think? But not exactly. According to astronomers "space" that we have know is 62 miles from the Earth's surface. This high point of 62 miles is known as"the Karman Line. The Dome is said to be 3,100 miles tall. Even though these weather balloons have soared more than 62 miles from the surface of Earth but they're not far from the claimed structure over.

What is the best way to get further than 3,100 miles? Rockets. There is online footage of rockets blasting into space using cameras attached that point downwards and showing the rotating Earth while the vehicle makes its way thousands of miles across the sky.

If the Dome was in existence this would indicate that rockets are launching in the sky, heading to a hidden place and "waiting for it to clear" until they have be able to get back home. NASA has launched more than 130 spaceships over the last five decades. What is the reason no one has seen one of these spacecrafts? Did anything ever crash in the Dome? How do Flat Earthers find out that there's a Dome? Did anyone ever see it? Have you ever seen it?

4. It is home to the Sun, Moon, planets and stars are within the Dome.

According to calculations of astronomers according to calculations, the Sun's diameter is 863,705 miles. The lunar diameter of 2,323 miles. Astronomers estimate it that the Moon is 235,000 miles to the Earth and that the Sun is approximately 93 million miles from the Earth.

But Flat Earthers think that the Sun as well as the Moon are of the same size. 32 miles in width. In other words we are aware the fact that 1.3 million Earths could be contained in the Sun. But flat Earthers think it is the Sun is only 20 percent smaller that London. According to the FE Model, these celestial bodies revolve about each other around the North Pole 3,000 miles above the Earth 100 miles below the Dome.

When we're in moon's light, it's nighttime. If we're in the sunlight of the Sun it's daylight. They also say that the sun's light is only light in a beam similar to the spotlight.

Meteorologist Keith Carson, disproved this in just a few seconds by pointing out

I) The Moon is not visible in this model during the daytime.

I) the hours of daylight will be identical all through the year.

III) You will always be visible to the Sun even at night.

In addition there is the possibility that the Sun changed its size during the day on a plane that was flat when it moved closer or further from your position. If you can't be able to see the Sun because it's shining down, it should be in a position to see the glowing ground.

In addition it is also the case that we have the same view of the Moon. It is simple to confirm as the craters that form in Moon's surface Moon are always the same. Why can't we observe the opposite face of the Moon on an FE Model?

The Sun is around 400 times larger that the Moon. This is because the Sun is also 400 times further to the Earth as compared to the Moon. This is the reason why both the Sun along with Moon appear to be the same size. Moon look of the same size on the night sky. Since the Moon is identical to the Sun in the sky and the Moon could completely obscure the Sun. How does an eclipse function on flat ground? Based on Flat Earthers they believe there are ANOTHER celestial body that lies below the Dome that blocks the Sun.

Flat Earthers cannot determine what this body is. Is it a moon? Are you referring to another star? Which celestial object is it composed of? Why can't we see it at another place? What's the explanation for the object that blocks the Sun during an eclipse (which is evidently the Moon) isn't actually the Moon? If someone must invent an unproven inexplicable object to support their conspiracy, they've entered the realm of the supernatural.

5. The Earth is enclosed by an Ice wall.

According to the Flat Earth Model, Antarctica isn't actually an actual continent. Instead, it's an uninterrupted 45-metre tall (150ft) glacier that extends across the globe.

If you've ever seen images from "The Ice Wall," it's a simple image of a floating, thick platform constructed of ice. These structures are called Ice shelves. They can be located within Antarctica, Greenland, Canada and Russia. There is nothing new about these structures.

How does the Ice Wall work according to Flat Earthers? There's no definitive answer. Why does it not melt? Is it because of the edge? Do you see something else? Different FE members give different answers. Many believe that they believe that the Ice Wall goes on forever. Some believe there's a continent that is forbidden, called Antichtone on the opposite side. Many believe the continent is covered by ice, each of which hides continents between them.

You may be thinking "Why is it that a small bunch of flat Earthers head to Antarctica? 37,000 people visit Antarctica each year. There aren't any Game of Thrones walls anywhere? How can that be possible If such a thing is even possible?"

The rumor is that tourists are allowed only to go into a tiny portion of Antarctica in order to be able to view this wall. You can't also explore Antarctica randomly. You'll require an guide. It is believed that tourists are guided by a guide, so that visitors don't wander off and stumble upon the famous Ice Wall.

In real life, escorts perform an additional task - to ensure that you don't get killed. Average temperatures in Antarctica is 49 degrees Celsius. It's triple the temperature of the temperature in your freezer. This is the average temperature! That's expected. The lowest temperatures are -89.6 degree Celsius. What is the time you are required to endure when you ventured into this temperature with no significant quantity of planning? Yet, conspiracy theorists think they have United Nation troops patrol the Ice Wall, ready to shoot anyone who comes close to it.

Let's pretend for a moment that this was the case. How many people could it take to construct the wall of 75,000 miles across? Even if there were an individual guard for every mile (which isn't effective in stopping people from entering,) that's 75,000 guards who must be trained and expect to be able to stand in the coldest climate anywhere on

Earth... for the rest of their lives. How can this be done?

The most important question regarding the Ice Wall is this - What is the conspiracy theory behind it? that it exists? Someone must have found it. Who can believe such a thing when nobody knows who did it, or what time? What is the proof that there are guards on the Ice Wall has guards if there aren't any photos of them, and nobody claims to ever observed them? This is one of the shady rumors that seem to be a bit naive when you consider it.

6. The concept of gravity isn't real.

When I first learned that Flat Earthers don't believe gravity, I asked the same question that most people ask "Then why do things drop?"

In the event that you dropped a football in an ocean, then why does the ball flounder? What is the reason gravity doesn't cause it to sink? If you're Flat Earthers The solution is the exact same: the buoyancy of the earth and its density.

But it seems that the FE community refuses to acknowledge the fact that buoyancy is dependent on gravity in order to exist. Take a flight with zero gravity and buoyancy ceases working precisely the time zero gravity takes over.

In addition, if gravity does not exist, what's the rationale behind why you can see the Sun and Moon appear to hover over the sky? Are you aware of is the Flat Earth explanation is? Simple. There's no such thing.

There are many ways to prove the existence of gravity but here is the most straightforward and most succinct way to prove it. Tablets and smartphones have sensors which detect gravity. This is why they adjust the screen display as it rotates.

Here you go. The concept of gravity was demonstrated in just two sentences. Sir Isaac Newton would be proud (if the scientist had known what the term "smartphone" meant.)

7. The Moon is self-illuminating.

If the Sun functions as an upward-pointing spotlight on the Flat Earth, then it must also point at the Moon to light it. Therefore the Moon is self-illuminating. This is not enough to explain why self-illumination could alter the quantity of Moon which is lit during the course of a 29.5-day cycle and what the source of power to this light source is the reason the craters are evident on the Moon's surface and how we are able to clearly discern shadows on the moon's surface.

8. The Earth is just about a couple of thousands of years old.

There are many different studies that catalogue thousands of different species of animals and plants that are half one billion years old. Additionally, the research of every archeologist, geologist volcanologist, seismologist and paleontologist agree with the idea that Earth has a lifespan of 4.5 billion years old.

What is the reason people believe that Flat Earth?

While The Flat Earth Society states that it is "no affiliation to any religions that are recognized as legitimate," their website uses at least 17 verses from the Bible which they claim to validate their beliefs, which includes the following:

The Earth is flat.

Did you know I was laying the foundations of the Earth?

Job 38:4

The trees grew huge and sturdy; its peak reached to the sky and was visible to all the extremes of the Earth

Daniel 4:11

And after these things I saw four angels standing on four corners of the Earth

Revelations 7:1

Water between the Sky and Heaven

And God said, Let there be a firmament in the midst of the waters, and let it divide the waters from the waters. And God made the firmament, and divided the waters which were under the firmament from the waters which were above the firmament: and it was so

Genesis 1:6-7

Praise Him, o highest heavens, and you waters above the skies

Psalm 148:4

The Earth Never Moves

Mine hand hath laid the foundation of the Earth Isaiah 48:13

Which shaketh the Earth out of her place, and the pillars thereof tremble

Job 9:6

When the Earth and all its people quake,

it is I who hold its pillars firm

Psalm 75:3

He set the Earth on its foundations, so that it should never be moved

Psalm 104:5

The world is firmly established; it cannot be moved

1 Chronicles 16:30

Referencing the Dome

Hast thou with him spread out the sky, which is strong, and as a molten looking glass

Job 37:18

Stars will fall to Earth

All the stars of Heaven will be dissolved. The skies will be rolled up like a scroll, and all their stars will fall like withered leaves on the vine and foliage on the fig tree

5 Isaiah 34:4

Immediately after the tribulation of those days, the Sun will be darkened, and the Moon will not give its light, the stars will from the sky, and the powers of the heavens will be shaken

Matthew 24:29

And stars of the sky fell to the Earth, like unripe figs dropping from a tree shaken by a great wind Revelation 6:13

Just because the Bible references a dome or the foundations of the Earth doesn't make these statements factual. Scripture isn't evidence, no matter how holy it is perceived. Evidence is independently verifiable. It was common knowledge that the Earth was round when the Bible was written. Although the Catholic Church did inaccurately believe that the Sun revolved around the Earth, a flat plane has never been a core belief.

So, if the writers of the Bible didn't believe in Flat Earth, why did they refer to "the foundations" and "the pillars of the Earth?"

It's simply an expression. The Bible has hundreds of metaphors that are not supposed to be taken seriously. Here are other passages that are clearly metaphorical -

The teaching of the wise is a fountain of life

Proverbs 13:14

Jesus said to them, 'I am the bread of life.'

John 6:35

The Lord is my rock

Psalm 18:2

I am the vine; you are the branches

John 15:5

The Lord is my shepherd

Psalms 23:1

No one believes God is a sheep-herder or Christ is a loaf of bread because these phrases are clearly not meant to be taken literally.

This is a perfect example of twisting the Bible's meaning to fit an agenda. Sometimes, they have to really clutch for straws. This quote, "Hast thou with him spread out the sky, which is strong, and as a molten looking glass" is supposed to refer to the Dome even though a dome isn't mentioned!

In case you are wondering, there are religious people besides Christians that believe we live on a flat plane.

Flat Earthers argue that since a Flat Earth has been a part of almost every main religion throughout history, there has to be something to it. But just because multiple religions shared a belief doesn't mean it's true. That's not evidence. That's gospel.

Ancient Egyptians often depicted their sky god, Geb, holding up the heavens while standing on a Flat Earth. Do Flat Earthers believe in Geb or any aspect of Ancient Egyptian mythology? Do they believe in the underworld or any of their gods?

In Norse myth, the flat realm of the gods, Asgard, was held up by Yggdrasil the World Tree. Do Flat Earthers believe in Yggdrasil? Or Thor? Loki? Do they believe in the Frost Giants that the Asgardians battled?

In Hindu beliefs, the Flat Earth lays upon the backs of elephants who are, in turn, standing on the shell of the World Turtle. Do you see my point? Although each of these religions had a fantastical mythology, Flat Earthers cherry-pick the one element that agrees with them.

This fallacy reminds me of George Bernard Shaw's quote, "No man ever believes that the Bible means what it says; he is always convinced that it says what he means."

The Flat Earth Society

As I explained in the introduction, the Flat Earth Movement had lain dormant until Samuel Rowbotham brought the idea back into the public conscious. Although his theories were ridiculed and his reputation was tarnished, there were several people who were intrigued by the possibility of a disk-shaped planet.

When Rowbotham died in 1884, his followers formed the Universal Zetetic Society to spread his message. Although the group made its way to Zion, Illinois, it collapsed in the 1940s when its promoters, John Dowie and Wilbur Voliva, passed away.

Although this conspiracy seemed to be a fad that had come-and-gone, its popularity rejuvenated in England in 1956 when Samuel Shenton created the International

Flat Earth Research Society in Dover, England. It was Shenton that concocted many core beliefs to dismiss the Round Earth concept. For example, Shenton was the first person to say photographs of a curved Earth were due to a wide-angle lens.

When Shenton died in 1971, Charles K. Johnson became the society's president. Johnson spread his information through a tabloid called Flat Earth News. It was Johnson who stated that the Sun and Moon are only 32 miles in diameter, which is a staple in Flat Earth beliefs. Within a few years, the society had a big enough impact for Johnson to have an interview with the Los Angeles Herald Examiner. In this discussion, Johnson emphasized that Science is a religion and Jesus Christ was a Flat Earther.

Although it appeared that Johnson's group was finally making waves, disaster struck. A fire destroyed Johnson's home as well as the group's library and archives, putting the society in jeopardy. When Johnson died in 2001, the group seemed to be on the way out.

However, the group was relaunched by Daniel Shenton (no relation to Samuel) in 2004 under the banner, The Flat Earth Society.

According to the website, theflatearthsociety.org, Flat Earthers' mission is "to promote and initiate discussion of Flat Earth theory as well as archive Flat Earth literature. Our forums act as a venue to encourage free thinking and debate."

They refer to society's blind devotion to science as "the Globularist lies of a new age."

Although the site gave a lot of insight on the origin and beliefs of Flat Earth, it was very inconsistent with updates until it was closed down in 2018. There were only six entries in the website's blog and it wasn't once updated after 2016. The grammatical errors were atrocious. In the article, Early Parallaxian Theory in a Nutshell, it stated that the Alexandrians believed "the sun be far away and massive."

The article, Exclusive Interview with Mark Sargent read, "Go to the beech (sic), look at a mountain, look at a weather balloon - the only people showing it round are NASA."

Not only are these spelling mistakes and grammatical errors embarrassing, they were never corrected. Also, these articles weren't written by a random member. They were submitted by the Secretary, John Davis, who has the highest position in the company apart from the President and the Vice President. It's very difficult to take the Society seriously, when the higher ups fail at fundamental sentence structure.

In 2013, a part of the Society broke away to form a web-based group called... The Flat Earth Society. That's right. For five years, there were two organizations with the exact same name... because Flat Earth wasn't already confusing enough.

This group is now considered to be the official Flat Earth Society. (To make it slightly less confusing, the first Flat Earth Society is initialized as FES and the second one is initialized as TFES.) TFES broke away from the original group as it believed some Flat Earth beliefs had become outdated (which is ironic on so many levels.)

Before I go any further, I must stress that not all Flat Earthers are members of the FES. Although that might seem contradictory, many people who were raised Christian may pray to God but they perceive the Catholic Church in a negative light. I just need to stress this as I do not want to imply that every Flat Earth believes what the society preaches.

At the time of this book's publication, TFES has almost 60,000 followers on Twitter. It's hard to say how many people actually believe in Flat Earth since many people follow TFES to mock them (including dozens of my friends.) As I said, many Flat Earthers don't agree with TFES so the number of people who believe in Flat Earth could be much higher or lower.

Although TFES ridicules many scientific fields like astronomy and physics, science has given society vaccines, cars, electricity, solar power, vehicles, the Internet, Wi-Fi, radiotherapy, chemotherapy, penicillin, surgery, engines, x-rays, anesthesia, phones, cameras, and an understanding of every facet of the universe.

On the other hand, The Flat Earth Society has contributed nothing. No inventions. No peer-reviewed theories. No verified studies. Since there has never been a single breakthrough in math or science from the TFES and their members reject the collective knowledge of the scientific community, it's impossible to take The Flat Earth Society seriously.

Characteristics of a Flat Earther

A YouGov survey polled 8,215 US adults and confirmed that 16% of them believed the Earth is flat. However, 34% of the millennials who were polled believe in a Flat Earth.

Flat Earthers (and conspiracy theorists in general) can be broken down in four categories –

i) The Persecuted see themselves as the victim of the conspiracy and try to warn others to avoid them falling victim to it.

ii) The Obsessed Enlighteners have supposedly discovered a plot e.g. an astronaut was about to come clean about the Moon landing being fake so he was murdered by NASA employers. When a Flat Earther uncovers this "truth" they want to tell everybody. Nowadays, Obsessed Enlighteners can spread their misinformation faster than ever due to social media or by self-publishing their "revelations."

iii) The Opinion Leaders can't separate their opinion from facts. They only pursue information, no matter how flimsy, to support their beliefs.

iv) The Witch Hunters are the most dangerous conspiracy theorists. They directly accuse people of being a part of the cabal. During the 16th and 17th century, it was these kinds of conspiracy theorists who were accusing innocent people of witchcraft, resulting in them being sentenced to death.

So, what's it like to talk to a Flat Earther? From my experience, they rely on one major tactic to convert others to their way of thinking – gaslighting. Gaslighting is when a person tries to make you question your own reality. If gaslighting sounds scary, it gets worse. These tactics are mainly used by abusers, dictators, narcissists, and cult leaders.

According to psychologist, Stephanie Sarkis, people who gaslight will do the following –

i) Gaslighters tell you lies. Many Flat Earthers state that visionary scientist, Nikola Tesla, believed the Earth is flat even though he didn't (and I will prove this later in the book.)

ii) Gaslighters deny that they said something contradictory even if you show them proof. If a Flat Earther says, "I never said _____" on Twitter and then you show them the tweet where they made this statement, they will still deny it.

iii) They wear you down. In this day and age, this is usually done with spamming messages and tweets.

iv) Gaslighters throw in positive reinforcement to confuse you. A Flat Earther might say that you seem so smart that they can't understand how you fell for The Globe Lie.

v) They project. Like most childish arguments online, the conversation will quickly descend into homophobia, sexism, racism, transphobia, and xenophobia. In at least two dozen arguments I have had with Flat Earthers, one of them said, "Bet you voted for Hillary, didn't you?" This is a little odd because I don't live in the United States and I am not American.

vi) Gaslighters try to align people against you. The FE community regularly state that NASA are Satan worshippers, Bill Nye is an idiot, Neil deGrasse Tyson is a shill, and every space facility works for the Illuminati.

vii) Gaslighters call you crazy. This is easy to do since any statement sounds impossible if it is oversimplified. Every day, I receive messages saying something like, "Ha! You believe we are stuck on a sphere spinning around a giant nuclear fireball. And you guys think WE are crazy!"

viii) Gaslighters tell you everyone is lying to you. As I am compiling this list, I am currently getting messages on Twitter saying how respected astronomers like Brian Cox and Kip Thorpe are shills paid by NASA to lie about the true shape of the Earth. The bizarre thing about this accusation is that they don't have any proof. If respected scientists like Cox or Nye or Tyson or Hawking were paid shills, how does the FE community know? Has anyone ever admitted to being a NASA shill? If they haven't, how do conspiracy theorists know that NASA shills exist? What are they basing this off? (For the record, I have been accused of being a NASA shill over 16 times... today.)

Another tactic that is regularly used by conspiracy theorists is peer pressure. The very first sentence on the blurb of Eric Dubay's book, The Flat-Earth Conspiracy, reads, "Wolves in sheep's clothing have pulled the wool over our eyes." Flat Earthers try to sound like you are stupid or you are missing out if you don't convert to their way of thinking. Here is a message I received from a Flat Earther who, at the time, I had never contacted -

"You have been indoctrinated since birth, my friend. It is why you and every other sheeple moron react in the exact same way when somebody questions your false paradigm. You are programmed to react in that way. You have no original thoughts and hate anyone that does."

Much like gaslighting, pressurizing potential converts is another popular tactic used by cult leaders. If a truth was self-evident, you wouldn't need to pressurize people to believe it. If a scientist tries to explain a theory, there's a good chance that the average Joe won't be able to understand it. When this occurs, the scientist simplifies the theory so it's easy to comprehend. If you told a scientist that you didn't understand his theory, he or she is not going to call you an idiot and say you are brainwashed. If a person doesn't understand what you are trying to say, it's up to you to find a way for them to comprehend it. Einstein famously said, "If you can't explain it simply, you don't understand it well enough."

When I first told my friends that Flat Earthers exist, most of them thought they are just "winding me up" and don't sincerely believe that our world is a flat plane.

Clinical neurologist and assistant professor at Yale, Steven Novella, observed many Flat Earthers and confirmed they are not "punking" others with their beliefs.

Upon further investigation, Novella had this to say, "In the end that is the core malfunction of the Flat-Earthers, and the modern populist rejection of expertise in general. It is a horrifically simplistic view of the world that ignores (partly out of ignorance, and partly out of motivated reasoning) to real complexities of our civilization. It is ultimately lazy, childish, and self-indulgent, resulting in a profound level of ignorance drowning in motivated reasoning."

Chapter 6: Flat Earthers Throughout History

If the Flat Earth concept is making a comeback, surely there are a handful of iconic figures who champion its cause, right?

Well, the most vocal FE celebrity is the rapper, B.O.B. He talks about his beliefs in interviews, Twitter, YouTube videos, and even in his lyrics. He set up a GoFundMe called Show BoB The Curve for $1 million to conduct an experiment to prove the Earth isn't curved. Personally, I'd assume it would be far cheaper just to pay for a pilot to fly him high enough to see the curve. After two months, he only received $6,872 from 227 people. Technically it was $5,872 from 226 people since B.O.B. was nice enough to pledge $1000…. to himself. This means that he didn't even receive 1% of his pledge. Although Flat Earthers claim "they are winning," it appears that only 0.00003% of humanity believe in the concept enough to stake their money on it. And B.O.B's the most well-known Flat Earther alive today.

You know who else Flat Earthers use to advance their cause? George Bernard Shaw.

You might be thinking, "Wait... the Nobel Prize winner and Oscar-winner, George Bernard Shaw? One of the most respected writers in Irish history? The author of Pygmalion which was adapted into My Fair Lady? He was a Flat Earther??!"

This idea comes from Shaw's quote, "In the Middle Ages, people believed that the Earth was flat, for which they had at least the evidence of their senses: we believe it to be round, not because as many as one per cent of us could give the physical reasons for so quaint a belief, but because modern science has convinced us that nothing that is obvious is true, and that everything that is magical, improbable, extraordinary, gigantic, microscopic, heartless, or outrageous is scientific."

However, Flat Earthers who use this speech as an argument to paste Shaw as one of their own fail to point out that this passage is said by a character in his play, St. Joan. Shaw had no belief whatsoever in Flat Earth.

In his own words, "I must not, by the way, be taken as implying that the Earth is flat, or that all or any of our amazing credulities are delusions or impostures."

Cherry-picking data and statistics to fit your argument is known as The Texas Sharpshooter Fallacy. Another example of this is a quote from the historian, Charles Fort. He said, "It was impossible for Columbus to prove that the Earth is round." Fort didn't believe in Flat Earth but this quote gives the impression that he did. As I have explained in the introduction, mankind knew centuries before Columbus' time that the Earth was round. Regardless, Charles Fort's quote was used on The Flat Earth Society's website.

Another quote on the website is from the visionary author, George Orwell. It reads, "Just why do we believe that the Earth is round? I am not speaking of the few thousand astronomers, geographers, and so forth who could give ocular proof... but of the ordinary newspaper-reading citizen, such as you or me."

Not only was Orwell perceived as one of the greatest writers to ever live, he was passionately against governments controlling the media; a concept that Flat Earthers adamantly believe in.

As you can probably tell, this quote wasn't suggesting that Orwell believed in Flat Earth. He was saying that the layman relies on the work and knowledge of experts to understand the more complex aspects of the world since it is impossible for the average man to study or understand every scientific principle thoroughly.

When asked directly about Flat Earth, Orwell had this to say, "As for the Flat Earth theory, I believe I could refute it. If you stand by the seashore on a clear day, you can see the masts and funnels of invisible ships passing along the horizons."

Although the Flat Earth Society use quotes from Charles Fort, George Bernard Shaw, and George Orwell to highlight their beliefs, none of these men believed in Flat Earth. So, who does believe in it?

Singer, Tila Tequila, is a part of the FE community. The 7th US president, Andrew Jackson passionately dismissed the curved Earth concept. Some sources state that NBA players, Shaquille O' Neal and Kyle Irving are Flat Earthers. Although they find the concept fascinating, they are aware that it is an impossibility.

However, none of the people mentioned above are scientists. If the Earth was flat, how come there isn't a single astronomer who has deduced it? How could the greatest minds miss such a rudimentary truth? Is there a single respected scientist who believed the Earth is flat?

This brings us to a Swiss physicist called Auguste Piccard who performed many experiments in the early 20th century. He studied the Earth's upper atmosphere, cosmic rays, and invented a deep-sea submarine called a bathyscaphe. He was the first person to view the Earth from ten miles up. The main character of Star Trek: The Next Generation, Jean-Luc Picard, was named after the renowned physicist.

However, Piccard is best known for believing the Earth is flat... even though he didn't. In the 1931 magazine, Popular Science, the quote "Earth seemed a flat disk with upturned edge," was attributed to Piccard even though the physicist adamantly knew that our world was curved. In fact, he repeatedly states the Earth is a globe in his 1956 work, Earthy, Sky and Sea –

i) "The geophysicists have installed a network of stations over the globe" – p120

ii) "Humanity will find enormous resources in the seas, which covers three-quarters of the globe." - p147

iii) "Because in ten years, perhaps, a brave explorer, yet un-known, will desire, on the other side of the globe, to make dangerous experiment." - p187

Piccard never said the Earth was flat. He said it appeared flat from his viewpoint. However, conspiracy theorists cherry-picked a single sentence of Piccard and mutilated its meaning for their own cause.

This isn't the only time that Flat Earthers have done this. Many Flat Earthers claim that Albert Einstein believed the Earth was flat. This is based on the following quote, "I have come to believe that the motion of the Earth cannot be detected by any optical experiment."

In case you're wondering, that's not a fake quote. Einstein did say that. However, it's only part of the quote. Flat Earthers tend to omit the final part, which completely changes the context. Einstein's full quote is, "I have come to believe that the motion of the Earth cannot be detected by any optical experiment, though the Earth is revolving around the Sun."

Why would a Flat Earther remove the last part if all they care about is the truth? Flat Earthers are guilty of only using part of a quote to suit their agenda. This technique is known as Quote Mining. It is often used in court cases and political debates.

This leads me to the most respected scientist in the eyes of Flat Earthers – Nikola Tesla. Tesla is unquestionably one of the most gifted scientists to ever live. He discovered x-rays, invented the rotating magnetic field, the Tesla coil, the induction motor, and the remote-control torpedo. He is one of the first people to work with solar powered technology and hydroelectricity. Tesla was smart enough to perform calculus in his head and he was the first person to hypothesize the possibility of Wi-Fi in the future. However, Tesla is best known for learning how to use alternating currents (AC) which advanced our ability to use electricity exponentially.

Because of his genius, it is difficult to contest Tesla in any scientific field. Which is why this quote attributed to Tesla is seen as the Flat Earth smoking gun, "Earth is a realm, it is not a planet. It is not an object, therefore, it has no edge. Earth would be more easily defined as a system environment. Earth is also a machine, it is a Tesla coil. The sun and moon are powered wirelessly with the electromagnetic field (the Aether). This field also suspends the celestial spheres with electromagnetic levitation. Electromagnetic levitation disproves gravity because the only force you need to counter is the electromagnetic force, not gravity.

Though free to think and act, we are held together, like the stars in the firmament, with ties inseparable."

Now, just because Tesla was a genius, it doesn't mean he can't be wrong. Tesla absolutely believed that he would live to 150 if he drank whiskey every day. (He died at 86.)

However, that's irrelevant because Tesla never said the above quote. The earliest recording of this quote is on January 11th 2016 on Facebook by Darrell Fox.

So, you would assume that Fox made up this quote and attributed it to Tesla, right?

Wrong. Fox never intended this quote to be attributed to the Serbian scientist. Fox admitted that "The first part is my quote. The second is Tesla."

So, the only part that can be attributed to Tesla is, "Though free to think and act, we are held together, like the stars in the firmament, with ties inseparable." When

Tesla refers to a "firmament," he didn't intend to be taken literally. We know this for a fact because this quote is taken from his 1900 article, The Problem of Increasing Human Energy with Special References to the harnessing of the Sun's Energy. In this article, Tesla refers to the Earth as a globe 12 times! Here are several examples -

i) This metal, it would seem, has an origin entirely different from that of the rest of the globe.

ii) It is a well-known fact that the interior portions of the globe are very hot.

iii) We should be enabled to get at any point of the globe a continuous supply of energy, day and night.

If you are worried that I'm taking quotes out of context, feel free to read the entire article online.

Because Darrell Fox coupled his piece with Tesla's, readers assumed the entire text was said by the famous inventor. Tesla couldn't believe in a flat world because his theory on wireless transmission relied on a round

Earth. Although Flat Earthers see Nikola Tesla as the poster-child for their beliefs, he knew the world was a globe.

I think historian, Klaus Anselm Vogel, put it best when he said, "No cosmographer worthy of note has called into question the sphericity of the Earth."

Everyone Would Have to Be in On It

If the world was flat, our understanding of every form of science would be wrong. This means that the building blocks of gravity, physics, geology, astronomy, and many other scientific fields were founded on a lie.

Flat Earthers emphasize that NASA faked the Moon landing. If that was true, that means hundreds of thousands of people who've worked there have been lying every day for over half a century.

But it's not just NASA. It's not just astronomers. Paleomagnetists study the Earth's magnetic field. This field is generated by molten metals within the Earth and it protects our planet from solar radiation. Without the magnetic field, everyone would be dead within minutes. Paleomagnetists study a Round Earth model to understand the planet's magnetic field. How can paleomagnetists have such a deep understanding of the magnetic field if they got the planet's shape wrong?

Seismologists study the lithosphere to find more accurate ways to predict earthquakes. Geo-dynamicists study tectonic plates. Volcanologists study a Round Earth model to understand how the planet's inner core, outer core, and magma chambers work. Surveyors take the planet's curvature into account when measuring large distances.

When Philip Scofield spoke to several Flat Earths on This Morning in 2018, Scofield pointed out that he saw the Earth's curve while he was on a Concorde. The Flat Earthers dismissed this, stating that the windows were curved, which gave the illusion that the Flat Earth was bending.

So why doesn't anything appear curved while the plane is still on the ground? Some Flat Earthers claim that the windows are made of high-altitude glass that appears "curved" when it reaches a certain height in the sky. This means that people who construct aircraft windows are also in on the conspiracy.

Now, there is a mild curve at the very bottom of the glass but that is to protect the plane from wind resistance. A curve causes the wind to slide into the glass rather than slam into it. Although Flat Earthers see this is a lame excuse, car windshields have the exact same curve at the bottom of the glass for the same reason.

Climatologists, geographers, navigators, pilots, and sailors rely on a deep understanding of how the planet works. Either they are all wrong but are still able to understand seismology, physics, geology, etc. or they are all "in on it."

Even bridge engineers would have to be in on it. The 1,300-meter-long Verrazano-Narrows Bridge connects Staten Island and Brooklyn. Although the two towers on this bridge were built perfectly vertical, they are 41 millimeters away from each other at the top than at the bottom as a result of the Earth's curvature.

The Ivanpah Solar Electric Generating System in California's Mojave Desert has over 300,000 computer-controlled mirrors covering 3,500 acres of land. They track the Sun and convert its solar power for over 140,000 houses. According to the truthers, Ivanpah are in on it too.

Because of Google Earth, you would assume Google has at least one satellite but they don't. Google use satellites built by other companies including Japan's MTSAT Satellite Augmentation System, India's GPS Aided GEO Augmented Navigation, and Europe's Geostationary Navigation Overlay Service. So not only would Google have to be a part of this cabal, every space facility who constructed these satellites would have to be in on the conspiracy.

As you may be able to surmise, Flat Earthers don't believe in climate change. The ones that do believe it are under the impression that climate change is created by something besides greenhouse gases. So, what do the experts think? Between November 2012 and December 2013, 2,259 peer-review articles about global warming were studied by 9,136 climatologists. Although these climatologists are experts in different fields - atmospheric science, glacier changes, economic and social impacts, only one of them did not believe in climate change after reading these articles. And guess what? None of them concluded that the Sun was 32 miles in diameter or the Earth is flat. If these statements were true, you'd think one of the 9,136 experts would've noticed.

You know who else would have to be in on this conspiracy? Nintendo. The Nintendo Wii U console has a magnetometer in its controller that can tell where the player is positioned based on readings from the magnetic field of the planet (but only on a Round Earth Model.)

Many people believe that the weakest argument is one that relies on a conspiracy. I disagree. The weakest argument is one that relies upon dozens and dozens of conspiracies to be true. This sort of conspiracy collapses if any one of these mini-conspiracies is proven to be false. If you can prove the Moon landing happened, the Earth can't be flat. If you can prove asteroids and comets are real, the Earth can't be flat. If you can prove the Sun is bigger than our planet, the Earth can't be flat. If you can prove our planet has a magnetic field, the Earth can't be flat. If you can prove gravity, the Earth can't be flat.

Thousands of astronomers throughout history separately drew the same conclusions about the Earth, the Moon, the Sun, and other celestial bodies. The Ancient Greeks used the Antikythera Mechanism to track the position and phase of the Moon and predict solar and lunar eclipses. Another dial showed the Sun's and the Moon's position on the zodiac cycle. It could track the movements of Venus, Mercury, Mars, and other celestial bodies.

Iraqi astronomers from 300 AD used calculus to work out the position of Jupiter with pinpoint precision. This was about 1,400 years before the Europeans were introduced to calculus.

10,000 years ago, the Scots built a lunar calendar in Warren Field made of 12 man-made pits, each corresponding with a different phase of the Moon perfectly.

Hipparchus of Nicea calculated the distance of the Moon from the Earth was 245,000 miles in the 2nd century BC. Although the average distance is 235,000 miles, it is still astounding that Hipparchus calculated this 1,700 years before the invention of the telescope. Hipparchus achieved this by observing the Moon's shadow during an eclipse.

Posidonius of Ancient Greece concluded the Earth's shape and size by observing the Canopus star. Muslim scholars had to figure out the shape of the Earth and the distance from the Sun to calculate the distance and direction from any given point on the planet so they can pray in the direction of Mecca.

Aryabhata of India, Ptolemy of Alexandria, Anania Shirakatsi of Armenia, Alfraganus of Persia, Bede of England, Ibn Hazm of Andalusia, Strabo of Ancient Greece, Biruni of Iran, Isidore of Seville, and Caliph of Baghdad separately concluded that we live on a globe.

In the words of Neil deGrasse Tyson, "One of the greatest features of science is that it doesn't matter where you were born, and it doesn't matter what the beliefs systems of your parents might have been; if you perform the same experiment that someone else did, at a different time and place, you'll get the same result."

If this diabolical plot has existed for decades, millions of people have taken part in this global scheme. However, a conspiracy of this size could never work. Oxford University physicist and biologist, David Grimes, devised an equation to show how long it would take for a collusion of this magnitude to become publicly known. Grimes based this equation on three conspiracies that turned out to be true –

i) The NSA surveyed society without their knowledge or authorization

ii) The Tuskegee syphilis experiment withheld a cure for a disease from African-Americans

iii) An FBI scandal where Dr. Frederic Whitehurst's misleading forensic tests led to the death of several innocent people.

Grimes used these instances to gauge the efficiency of a conspirator's ability to keep a secret. He then looked at four prevalent conspiracy theories –

i) NASA never went to the Moon

ii) Climate change isn't real

iii) Vaccinations cause autism

iv) Pharmacy companies have the cure for cancer but won't release it so they can make more money from their products.

Grimes' formula considers the number of conspirators, how much time has passed, and the probability of a whistler-blower.

Since 411,000 people worked for NASA in 1965, Grimes worked out it would take 3.68 years before it became common knowledge that the Moon landing was fake. With 405,000 people working in climatology, the conspiracy would crumble in 3.7 years. The truth about vaccination would come out after 3.15 years. The cancer cover-up would be unveiled after 3.17 years.

The reason why the cancer conspiracy falls apart so fast is because there are more people involved. According to Grimes, 714,000 people work with pharmaceuticals. That's 714,000 people who have to keep mum about one of the biggest conspiracies imaginable. By this logic, the less people involved in a conspiracy, the longer it would take for it to be exposed. Grimes stated that if a hoax was to last five years, there can't be more than 2,521 people protecting it. For a hoax to last 25 years, there needs to be a maximum of 502. For it to last a century, you can have 125 people at most. With the amount of people allegedly involved in the Round Earth plot, the chances of it staying secret are zero.

Are Flat Earthers Qualified?

If you fall ill, you go to a doctor. You see a doctor when you feel sick because he or she is expected to be an expert on the human body.

You see a chiropractor if you pull a muscle because the chiropractor is expected to be an expert on fixing musculoskeletal problems.

You see an optometrist if you have an eye problem. You see a dentist if you have a toothache.

Each of these experts studied for years for their qualifications. Although astronomers also needed years of research to be considered an expert in their field, they are met with scorn and suspicion by Flat Earthers. Even though renowned astrophysicists like Bill Nye, Brian Cox, and Neil deGrasse Tyson effortlessly prove the Earth's shape with their decades of knowledge, they are bombarded every day on Twitter from Flat Earthers who claim to have disproven centuries' worth of science because they made some observations while looking at the Sun with their camera phone. That's not a joke. There are countless videos on YouTube where it appears that the clouds are behind the Sun, which Flat Earthers use to push their agenda. As photographer, Mick West said, this is obviously an illusion since the Sun is 93 million miles away. Sections of clouds are transparent so it appears that the Sun shines through it. Although any professional photographer would give the same answer, Flat Earthers dismiss this.

See, this is a huge problem with Flat Earther's certainty in their theory. These conspiracy theorists expects to know more about astronomy than an astronomer. They claim to know more about photography than a photographer. They believe they know more about physics than a physicist. It is scientifically ignorant for a person to have never studied astronomy, physics, chemistry, or geology, nor have they read any of the tens of thousands of peer-reviewed scientific papers describing the fundamental principles but expect to know more than any expert.

You can't just read some articles online, watch a few YouTube videos of space, make a handful of undeveloped observations about the Sun and Moon, and believe you have cracked a conspiracy that disproves the foundation that science is built on.

Flat Earthers reject this argument, saying they prefer to do their own research and make their own observations rather than regurgitating what so-called experts say.

We don't blindly follow astronomers. We listen to them because they are respectable figures due to their decades of experience. Neil deGrasse Tyson is popular because he is charismatic, charming, and funny. But he's also a student of Harvard, a scientist for Princeton, the director of the Hayden Planetarium, and the founder of the Department of Astrophysics.

Society listens to Carl Sagan when he talks about astronomy because he promoted SETI (Search for Extra-Terrestrial Intelligence) and chaired the Voyager records, which captured many photographs of the Solar System. Society listens to Kip Thorne talk about gravitational physics because he is a Nobel Prize-winning physicist.

However, society does not listen to Flat Earthers because most of their information comes from a weak understanding of science. If Flat Earth was hypothesized by a respected astronomer, maybe it would be given a little weight. However, this can never happen as any respected astronomer knows that the concept of Flat Earth is utterly divorced from reality.

The Moon Landing "Hoax"

Nearly every Flat Earther I spoke to believes in a myriad of conspiracies. I don't mean the classics like vaccinations, 9/11, climate change, etc. What I mean is that they believe in conspiracies that are… nuts. Flat Earthers have told me that the Vatican is hiding the remains of centaurs, mermaids, fairies, and vampires. They have told me that North Korea doesn't exist even after I have shown them pictures of me in

Pyongyang. They have told me that "The big lie is the fake aliens who are really the demons of the Nephilim possessing cloned bodies & the Flat Earth." I found that so ridiculous, I asked the Flat Earther to write it down for me. I have looked at it a hundred times and I still don't know what it is means. At the Flat Earth Convention in Birmingham, they gave out fluoride-free toothpaste because the FE community think that government puts fluoride in toothpaste to make people more subservient.

Despite these outlandish theories, I don't want to focus on them as it's lazy to say, "This person believes in mermaids and vampires so they must be wrong about everything else."

But there is one exception – The Moon landing. Since Flat Earthers believe that a Dome covers the entire planet, it is impossible to reach space. So not only do Flat Earthers refuse to believe that man walked on the Moon, they believe it is impossible.

Since they see outer space as a fantasy, they are under the impression that NASA was created to pocket billions of dollars. The organization claims that it cost $25.4 billion to put a man on the Moon. How much do you think it would cost to fake a Moon landing?

Way more. If faked, NASA still hired thousands of physicists, mathematicians, and engineers. They performed countless rocket tests. Fake or not, all of that would still cost billions of dollars. If the Moon landing was falsified, that means NASA is still paying for it since they have had to bribe workers to stay silent. They would have to monitor any information that could blow the whistle on the whole operation. NASA would have to keep this up forever.

It seems silly that a conspiracy theorist believes that NASA performed countless tests and spent billions of dollars on rockets, knowing that they could never go to space.

However, let's give the conspiracy theorists the benefit of a doubt. Maybe NASA had every intention of going to the Moon but they realized too late that it was impossible. To avoid being humiliated after spending a fortune on the project, they paid film director, Stanley Kubrick, to shoot a Moon scene in a closed set and hope the public would be none the wiser.

Many conspiracy theorists have pointed out inconsistencies in the photographs and videos of Neil Armstrong and Buzz Aldrin walking on the Moon. Unsurprisingly, none of the accusers are photographers or filmmakers since they don't have the slightest understanding of lighting, video, or perspective.

One common argument is this - How come there are no visible stars in any of the Apollo photos? I'm pretty sure if NASA was going to fake a Moon landing, they would remember that there are stars in space.

We can't see stars in this photograph because the camera was set to expose for broad daylight. If the photo was set so the stars were visible, the rest of the picture would have been overexposed.

Another photo of the Moon landing that conspiracy theorists use to discredit NASA is the one that shows the letter C on a rock, proving it's a prop.

However, there are photos of this rock in the exact same position without the C appearing on it. That "C" is a hair dropped on the film while the photo was being retouched.

Although conspiracy theorists dismiss this explanation, here's a more interesting point – Why wouldn't NASA just use AN ACTUAL ROCK?! Why make a fake one when there are rocks literally everywhere?

Most of the "inconsistencies" in these photos have straightforward explanations. For example, why does the American flag stay up even though there is very little gravity on the Moon?

The flag was designed to extend outward. Why bring a US flag to the Moon if it wasn't devised to billow out?

Another argument is how the astronauts look like they are weightless because the camera was slowed down, which gave the illusion that they were floating.

Ironically, this is the most compelling evidence to prove that the Moon landing did happen. Neil Armstrong and Buzz Aldrin walked on the Moon for 143 minutes. This footage was shown in its entirety when they landed on July 21st 1969.

At the time, the Apex disc recorder was used for slow-motion shots. This camera records footage at one-third the speed to highlight movement, especially during sport games. The footage on the Moon was shot at 11 frames per second. If the footage was slowed-down, that means we have to watch it at 33fps to see the astronauts moving normally. Since the footage was 143 minutes, that means that just under 48 minutes of film would be needed to slow down the images.

However, this idea falls apart since the Apex disk recorder could only hold 30 seconds of material at the time of the Apollo 11 mission. Because of this, that the footage was almost a hundred times too long to be filmed on that type of recorder.

The later Moon landings were recorded with NTSC video cameras which shot at 30fps. Since it was filming at triple the framerate, it would've been three times harder to fake.

Belief in conspiracies creates a domino effect. If you are convinced about one conspiracy being true, you are susceptible to believe others. Luckily, this works both ways. A person may lose belief in a conspiracy by being convinced that another conspiracy isn't true. The FE community cannot explain how Apollo recorded the Moon landing without using an Apex disc recorder. This means that the Moon landing happened, which means that space is real, which means the Earth is round. If a conspiracy theorist is convinced that one of their beliefs is false, it is likely that they will doubt the rest.

Chapter 7: Cameras Don't Lie

(Except When They Do)

There is one type of camera that Flat Earthers swear by – Nikon Coolpix P900. This camera is very popular because it has an astounding zoom with a ratio of 83. Flat Earthers act as if this camera is like the magical sunglasses in the film, They Live, that allow the wearer to see the world for what it really is.

While zooming in with this camera while looking at Mars or Venus, they will not look like solid objects. Instead, they will look like shimmering blobs. This is how Flat Earthers believe these celestial bodies really look. According to The Flat Earth Society, Mars, Venus, and Jupiter aren't really "planets." In fact, planets don't exist. Neither do moons, stars, or even galaxies. The TFES state that all these celestial bodies are "stuck" to the Dome and are not solid objects.

However, there's one enormous issue they don't realize. The reason why Venus looks so bizarre in this picture is because the P900 is missing a feature that every telescope possesses – manual focus. Since these cameras were never designed for stargazing, planets and stars appear as whirling globs when zoomed in.

As cited in the introduction, the first record of the Earth being proven to be a globe is when Anaximander saw boats disappear from the bottom-up as they sailed away. Now that cameras have zooming-in capabilities, this is more obvious than ever.

What is the Flat Earthers' explanation? Perspective. Conspiracy theorists act like "Perspective" is a magic word that can invalidate any scientific argument.

In Gordon S. Brooks book, The Earth Is Not Flat, the author states how perspective isn't a scientific concept. With Brooks' 50 years of experience as a photographer, he explains how the principle behind perspective is how things look smaller when they are further away from the observer. He explains it as thus, "There are two kinds of perspective: linear and aerial. Aerial perspective has to do with the way atmospheric distortion affects how we see things in the distance, and how using different colours and levels of detail can create this effect in a work of art. That's not what the Flat Earthers are talking about. They're talking about linear perspective, which you learned about in grade school by drawing things like railroad tracks and square buildings that reached for a distant vanishing point."

What's the Flat Earther's counter-argument? "The camera never lies, right?"

However, cameras weren't designed to gaze into space. That's what telescopes are for. At this very moment, the ELT (Extremely Large Telescope) is under construction in Chile. When it is completed in 2024, this 39-meter-diameter mirror will be able to take photos 16 times sharper than that of Hubble thanks to its adaptive optics. Through the ELT, we will see that Venus, Mars, and other celestial bodies are more than just blobby masses stuck to the Dome's firmament.

The Plane Paradox

Flat Earthers aren't a big fan of pilots. A pilot's existence jeopardizes the entire FE

Model. According to the FE community, every pilot in history knows that the Earth is flat. Somehow, every one of them has managed to remain silent about this Earth-shattering secret for decades.

But it's more complicated than that. Not only would every pilot have to be "in on it" for this conspiracy to work but some flight routes don't make any sense on the FE Model.

A flight from Santiago, Chile to Sydney, Australia takes 13 hours and 22 minutes. On a Flat Earth, the same flight takes 25 hours and 30 minutes. If you have ever taken this flight, your existence proves the Earth is a globe.

But that's not the end of it. This 25-and-a-half-hour flight is impossible since the longest nonstop flight in the world for a commercial airline, which was achieved by Singapore Airlines Flight 21 on October 11th 2018, was 19 hours. So not only would this flight be and extra 12 hours... it's not even feasible unless it stopped to refuel.

Now let's look at it from another point of view. Flat Earthers believe the Sun is only 3,000 miles above the sky. Commercial planes cruise at an altitude between 36,000ft to 40,000ft. Commercial planes never go higher than 45,000ft. That's just over 8.5 miles high. If the Sun was only a few thousand miles above the Earth, it would always appear above your viewpoint if you were in a plane. So why are there pictures where the Sun appears to be below the plane if it is at least 2,991.5 miles above it?

Another question that the FE community repeatedly ask is, "Why don't airplanes point their nose down to go over the curvature?"

When a plane is landing, its nose will be pointing upwards while it is going downwards. The angle of a plane's body only matters in so much as the wings are attached to it – but it is the angle and construction of the wings against the airflow that matter, not the body.

How does this work? The wings and the airflow produce lift. If the lift is greater than the weight, the plane goes up. If the lift is less than the weight, the plane goes down. To "follow the curve," a plane simply flies at an angle and speed to remain at the same altitude. As the Earth curves, so does the plane's path because it is maintaining a constant altitude above the curved surface.

That may be a little technical for some readers so let's look at it from a different angle. What if you could fly so high, you could see the round Earth with your own two eyes? Although you can't do this on a commercial flight, the curve can be observed on a MiGFlug jet.

MiGFlug is a Swiss company that allows clients to fly a fighter jet. I don't mean you get to just fly in a fighter jet. YOU get to fly it! For approximately $2,500, you can fly high enough to see the curve of the Earth. The company has been so successful, MiGFlug have built facilities in Florida, California, New York, England, Italy, Canada, Russia, Switzerland, Germany, Latvia, Slovakia, and Czech Republic. You can record the entire experience on a video camera.

What is the Flat Earth explanation for this? Curved windows. The whole thing is staged. The "customers" are paid to keep quiet when they see the world is flat.

If you are a Flat Earther, why wouldn't you want to verify it? If you can't trust the testimony of every single customer that MiGFlug has had since the company started in 2004, why not take to the skies yourself? Accessibility isn't a problem. Their services are available in 11 countries. You might think that nearly £2,000 is a bit expensive but I believe that's a very small amount of money to spend if the end result is learning if the greatest conspiracy in the history of mankind is true or not. If you think educating yourself is expensive, it can't compare to the price of ignorance.

Money,

So Much Money

If you irrefutably proved that the Earth was not round, you would become one of the most influential people in history. If NASA told you to keep quiet about it, would you?

I wouldn't. The revelation would shake the (flat) planet to its core (if it has one.) It would change the foundation of science. Why would I keep quiet about that? Why would anyone?

So, what would NASA do to maintain the fiction? Bribe you? How much money would it take for you to stay silent about the greatest scientific discovery and the biggest secret in history? A million dollars? More? Just to remind you, many scientists work a ruthless number of hours in the name of research and are usually underpaid and unappreciated. People don't become scientists for the money. They do it to understand the world and unlock the mysteries of the universe.

What would NASA do if bribery didn't work? Well, there's only one place to go from there. Threats. Keep quiet about Flat Earth or you will be killed.

But even this wouldn't work. When Giordano Bruno persisted that the Earth revolved around the Sun, the Roman Inquisition imprisoned him in 1593. Bruno was told he would be killed if he did not renounce his claim. Even though he was beaten and starved for seven years, Bruno stood by his belief. He was burned at the stake on February 17th 1600. It would only take one single person to refuse to be bullied by NASA to make the conspiracy public knowledge.

If Flat Earth is real, how is NASA able to maintain it? Are they giving out million-dollar hand-outs to everyone who stumbles upon the truth? Wouldn't paying off hundreds of thousands of people cause a bit of a red flag with the tax authorities? Have NASA murdered people for learning about the Ice Wall and the Dome? Wouldn't that leave a paper trail?

Why would NASA hide the fact that world is flat? To pocket money? They would be bribing so many people that it would be pretty redundant.

Also, NASA will pay volunteers $15,000 to lie in a bed for 24 hours a day for 90 days to measure the effects that zero gravity has on a person's body. What is the purpose of this if space isn't real? If they're pocketing money, why would they pay thousands of dollars for a pointless experiment?

If you want to get involved in astronomy for the money, I'm afraid you picked the wrong industry.

Antarctica's Elusive Ice Wall

Antarctica experiences one sunrise and sunset per year. Although there are several Flat Earth models, none of them can explain how a sunrise is an annual event in the South Pole. How can there be a visible Sun at midnight at both polar regions at the same time? There are at least 15 Antarctic webcams that show 24-hour daylight. If Antarctica surrounds the world as the conspiracy theorists claim, how can there be sunlight on all sides of it at the same time?

Why would scientists and Antarctic explorers lie about the frozen continent? Well, Flat Earthers believe the world is covered by an impenetrable Dome. A dome is shaped like a half-circle, which means it must have a bend. The Dome's arc would be most noticeable at the edge, which is the South Pole. Flat Earthers point out that there are encyclopaedias, including the 1958 book, Encyclopaedia Americana Vol II, that refer to the Dome being visible in Antarctica.

What is the "Ballers" explanation? Are these texts fake? No. The answer is much simpler. A "dome" is a type of glacier. It's a piece of upstanding ice that can be as high as 3,000 meters. There are dozens of these glaciers in Antarctica including Dome Argus, Holman Dome, Law Dome, and Dome Charlie.

Misinterpreting what a "dome" means is important because it shows that conspiracy theories tend to commit eisegesis. Eisegeis is when a person expresses their interpretation on a matter rather than the facts. Exegesis is the opposite; a truth based on undeniable solid information. Flat Earthers believe that the world is covered in a Dome and surrounded by an Ice Wall, which "Ballers" mistake for the continent of Antarctica. Flat Earthers read an encyclopedia with the words "Dome" and

"Antarctica" and that's all the verification they need.

Flat Earthers also refuse to accept that it is possible to travel from pole to pole even though this has been accomplished by many explorers. Monty Python actor, Michael Palin, achieved this feat and recorded his journey in the documentary, Pole to Pole.

For the record, Antarctica isn't sealed off. It's a tourist attraction. How come not one single Flat Earther has travelled to Antarctica to find the Ice Wall? You can book a flight to the White Continent with travelling companies like One Ocean Expedition or Quark Expeditions. Hundreds of explorers have walked, driven, and sledded through Antarctica.

You might think, "Wait, you can walk through Antarctica? Isn't it 2,000 miles across? Yes. Yes, it is. Sir Ranulph Fiennes, who is dubbed the world's greatest explorer according to the Guinness Book of Records, was the first person to trek from one side of the continent to the other on foot. It took six months for Fiennes to travel from Novo in the north to the Ross Sea in the south. Several people have accomplished the same journey since. However, none of them saw an Ice Wall that circled the Earth. If such a thing existed, this journey would've have been impossible.

How Would A Flat Earth Work?

If the Earth was flat, we would all be dead. That's a bit of a problem. All the heat on a spherical planet is concentrated in the core. If the Earth was flat, that concentrated heat, pressure, and radioactivity would be redirected to the planet's surface, incinerating everything. Also, the planet would be unable to maintain a magnetic core, which would destabilise the ozone layer. This would leave the Earth exposed to the Sun's rays, burning our world to a crisp. So, there you go. Flat Earth leads to everyone burning to death. Twice.

But that's no fun so let's say that wouldn't happen just so we can look at some other factors.

If the Earth was flat, there would be no axial tilt. No tilt means no seasons. Since the planet would be exposed to the same sunlight throughout the year, we would experience the same season all the time. Since weathermen and climatologists would have nothing to predict, they would be out of a job.

If the Earth isn't spinning, there would be no spinning motion to create hurricanes. If you see a F5 hurricane tearing through your neighbourhood, you can rest easy knowing that the Earth is definitely round.

On a Flat Earth, the human eye could see a candle flame at night time if it was 30 miles away.

According to Michael Stevens' YouTube channel, Vsauce, the gravity in the centre of a Flat Earth would be relatively normal. However, it would become warped at the edges. If a person in the centre ran towards the edge, it would feel like "they were fighting to climb up a steeper and steeper hill" as the gravity intensifies at the edge.

Stevens points out that a flat planet is impossible since, "Anything as massive as the Earth, shaped like a flat disk, would, under its own gravity, naturally collapse back into a ball. This is why in outer space everything more than a few hundred miles in diameter is round." So, not only is the Earth round, it's impossible for any large celestial body to be any other shape.

Bees Know the Earth Is Round

Bees communicate with each other by performing a dance known as "waggling." By waggling, a bee can inform another bee of a location to find food, a new home, flowers to pollinate, etc. The remarkable thing is that bees calibrate the position of the Sun as it moves around the Earth while performing its waggle.

This information was revealed during the Schafberg Experiment when scientists placed food for a swarm of bees on the other side of a mountain. Since the mountain was too high to fly over, the bees communicated with each other the angle around the mountain relative to their position, even though the insects had never flown in the area before.

When a bee needs to alert others of a good food source, it will waggle on a comb surface. For this dance, the insect turns in circles and bisects the circle at an angle upon every revolution. This bisection represents the Sun's position.

Let's look at this circle like a clock. If the bee bisected his circle at 6 o' clock, it's telling the other bee to fly towards the Sun to find food. 3 o' clock means the bee should fly to the left of the Sun. Bisecting at 12 o' clock means the bee should fly in the opposite direction of the Sun. This dance can take hours, which can be a problem since the Sun constantly moves.

But don't worry. The clever bees take that into account. While waggling, the bee calculates the change in angle based on where the Sun is now with the position of the hive considered. Bees can perform this type of waggle at night, which means they can calculate the Sun's position even when it's on the other side of the planet. It can take so long to reach the food, the bee adjusts their path based on where the Sun is to where it was when they were given directions. (Reminder – The bee's brain is the size of a seed.)

On top of that, bees can communicate how far the food is by how they wiggle their abdomen while bisecting the circle. The more they wiggle, the farther the distance.

When I explained this to a Flat Earther, the first thing he said was, "The reason all the bees are dying out is because the government is spreading poisonous chemtrails." First off, that doesn't have anything to do with the Earth's shape. I then explained how the bee population is the highest it's been in 20 years and the reason there are reports of bees dying out is because seven species of bee have recently become endangered. The other 3,993 species of bees are fine.

The Flat Earther retorted by saying, "So, they're not going extinct, proving you can't trust scientists. They lie about everything."

…Sometimes, you just can't win.

The Enigma That Is Gravity

Flat Earthers have a particular dislike of gravity since they are vociferously adamant that it does not exist. How does the tide work if it's not effected by the gravity of the Moon? FE founder, Samuel Rowbotham, said the tides are caused "from the rising and falling of the floating Earth in the waters of the 'great deep.'" This explanation is so flimsy, it is not even accepted by modern followers of Rowbotham.

So, if gravity doesn't exist, what replaces it? After all, things still fall down when you drop them. So why do they drop? The usual answer is 'density and buoyancy'. This is simply not true. But before we see why, lets remind ourselves how gravity works

Everyone has weighed themselves at some point. But what is weight? You may think it's the amount of stuff that you are made from – bones, the brain, kidneys, skin, eyes etc., but that isn't what you are measuring when you weigh yourself.

When you stand on weighing scales, you are compressing a spring, which tries to push you back up. Once the force of your body pressing down on the spring matches the force of the spring pushing you back up, you stop sinking and the indicator displays your weight by showing how far the spring has been compressed. So, weighing scales are measuring the force that you push downwards with. Where does this force come from? Gravity. Your weight is the force of gravity between you and the Earth.

Your weight is not the amount of "stuff" you are made of. Scientists have a different name for that, which is 'mass.' Mass is what you measure in kilograms or pounds. Weight is measured in units called Newtons. So, you might ask, "Why don't people say they weigh 750 Newtons?" The reason why is because your weight can be different depending on where you are. On the Moon, you weigh less because the force of gravity depends on your mass and the mass of the Moon. As the Earth has a much higher mass than the Moon, your weight on Earth is higher, despite your mass not changing. Since mankind have been weighing things before we understood how gravity works, people say "weight" when they should be saying "mass."

There is a force called gravity which pulls things together, and the strength of that force depends on the mass of each, and the distance between them when that distance is squared.

This means that not only does the Earth pull you down when you jump up, but you pull the Earth up too! Because the Earth has a much much much much much much higher mass than you, it doesn't move.

So how does buoyancy factor into this? Buoyancy is basically what makes things float. However, buoyancy doesn't pull things down, which is an obvious problem right from the start.

Basically, the buoyancy of an object depends on there being a difference in weight between the object and the same volume of water. But we already know that gravity produces weight, so buoyancy simply cannot replace gravity. In fact, buoyancy can't exist without gravity.

Sadly, the FE community refuse to acknowledge that gravity and buoyancy can co-exist. They refuse to accept that more than one force can work on the same object at the same time. For example, a hot air balloon is very heavy but it rises in the air because its buoyancy is greater than its force of gravity.

Although Newton didn't know what caused gravity, that in no way alters how correct he was. Newton's formulae are perfectly sound in 99.99% of calculations revolving around gravity. It took Einstein to come up with an explanation of how gravity comes about with his famous Theory of General Relativity. Despite the fact that Einstein's theory never worked off the back of Newton's, the formulae are almost identical. Although Einstein's formulae are more precise, we can use Newton's because the numbers that come out are the same.

While The Theory of General Relativity is over a hundred years old, it has never been proven wrong once. From the smallest laboratory benchtop experiment, to detecting gravitational waves, Einstein has been accurate every single time.

Some Flat Earthers understand that density and buoyancy cannot be used in place of gravity, so they say it is either magnetism or some form of static electricity. The obvious problem with this theory is that not everything is magnetic.

The static electricity idea is even worse. If that were true, we would be getting continual electric shocks every time we moved. Also, jumping up and down would probably kill you.

Gravity is just that – gravity. It is a feature of our universe, affecting everything within the universe, and the universe itself. Become a scientist and weigh yourself – you are measuring gravity!

Chapter 8: The Stars

One of the best open-and-shut cases to show that we live on a globe is to simply look at the night sky. The stars spin in the opposite direction in the northern hemisphere and southern hemisphere. This doesn't make any sense on a FE Model.

However, there is more than one way to prove this. One question that many Flat Earthers ask is, "If the Earth is spinning at 1,040mph as it revolves around the Sun at 67,000mph, while the Sun shoots through space at 450,000mph, why haven't the constellations changed in thousands of years?"

That sounds like a really good point... if it was true. The constellations do change. Earth's axis of rotation wobbles slightly – this is called precession. Although this wobble is very slight, it will affect how we see the stars over the next thousand years. For example, Polaris is known as the North Star. Although it is the most well-known star in the night sky, there's nothing particularly special about it. It is simply a bright star. It's not even the brightest star in the night sky. That honour belongs to Sirius.

Although Polaris is referred to as the North Star, it didn't use to be. Thousands of years ago, the Ancient Egyptians saw Thuban as their North Star because it was in the same position that Polaris is in now. In 11,000 years, Vega will become Earth's North Star.

Even an amateur astronomer can notice that Polaris has moved slightly in the last few decades. Anyone owning a telescope with an equatorial mount will have to adjust it for the movement of Polaris due to precession. So, it has noticeably moved within our lifetime, which further disproves the idea that constellations' positions are set.

Also, some stars (and constellations in general) are only visible from one hemisphere or the other which doesn't work on a Flat Earth. Not only that, but some constellations are only visible from anywhere at certain times of the year. The constellation of Orion cannot be seen between May and July except from Antarctica.

The Mind of a Conspiracy Theorist

Merriam-Webster's dictionary defines a "conspiracy theory" as "a theory that explains an event or set of circumstances as the result of a secret plot by usually powerful conspirators."

Although it's easy to assume that a conspiracy theorist is uneducated, it's more complicated than that. In 2014, political scientists, Joseph E. Uscinski and Joseph M. Parent, compiled a study on conspiracy theories and confirmed that it "cut across gender, age, race, income, political affiliation, education level, and occupational status." Although there are many educated Flat Earthers (some of them even have PhDs,) Uscinski and Parent's research showed that 42% of the conspiracy theorists didn't complete high school and only 23% graduated from college.

Uscinksi said that "Conspiracy theories are for losers," but not in the way you think. He means conspiracy theorists are "people who have lost an election, money, or influence looking for something to explain that loss." Dr. Thomas Swan, who studied cognition at Queen's University in Belfast and researches psychological traits and disorders, stated that most conspiracy theorists have a natural distrust or dislike towards authority. People like this believe that the US government knew about 9/11. The CIA had John F. Kennedy killed. NASA know the Earth is flat.

Although there are over 70 space facilities, NASA is the evil puppet master according to the FE community (probably because it's the only one they can name.) Dr. Swan's study noted that many conspiracy theorists have had a traumatic incident with an authority figure, be it a parent, teacher, or employer.

This inbuilt mistrust of authority transmogrifies into paranoia. Several Flat Earthers I spoke to were adamant that NASA had a file on them because they were getting close to revealing The Great Truth to the world.

Some of them blocked me because they believed I was a NASA insider. They even added my name to a list on Twitter called NASA Shills. If they are not accusing you of being a shill or a troll, they will mock your age, appearance, gender, or ethnicity. One Flat Earther attempted to defame me by telling others that I sent him death threats via Twitter. I told him I would delete my account if he could show a screenshot of these alleged threats. He blocked me seconds later.

These same people claim that they are Woke and only seek the truth. But the second Your Truth conflicts with Their Truth, they send abuse or try to have you reported.

Although this type of aggressive behaviour is unjustified, it is understandable. People are more willing to believe in conspiracies when they feel powerless and insignificant. For many conspiracy theorists like Flat Earthers, their belief is a safe haven. It eases their minds a bit by putting the Earth back into the centre of the universe. For millennia, we believed the universe was made for us. In recent years, humanity has realised how mind-bogglingly big the universe is. Astronomers have calculated that the universe is 92 billion light years across. That's 542. 8 septillion miles. That number looks like this –

542, 800,000,000,000,000,000,000,000 miles.

We could have never anticipated how gargantuan the cosmos are nor could we ever imagine how small our planet is by comparison. This poses a huge problem to religious people, especially Creationists. Why would God make such a big universe if only one planet was important… unless the universe isn't as big as the "experts" claim. By doing so, the conspiracy theorists place themselves, their views and needs, back into the spotlight.

You don't have to lie about the Earth's shape to feel special. Just because the Earth is one of countless planets in the cosmos doesn't make it any less incredible. Many astronauts who have viewed the Earth from orbit experienced a cognitive shift in awareness. This is known as the Overview Effect. By gazing at our fragile world hanging in space, astronauts have this incredible sensation where they rethink their perception of reality. Despite humanity's borders, politics, ideologies, religions, history, and beliefs, we are all human and Earth is the only home we have ever known. From space, viewing our planet creates a need to form a peaceful society on a planetary level. Many astronauts including Chris Hadfield, Scott Kelly, Tom Jones, and James Irwin said that they experienced the Overview Effect. They believe that peace can only become possible with a better understanding of the world and beyond it. This is impossible if conspiracy theorists obsess about long-disproven pseudoscience.

To accept a conspiracy like Flat Earth despite its glaring holes, gullibility has to be a factor.

Let me explain. When the 1999 film, The Blair Witch Project, came out, it took the world by storm because it was the first mainstream found-footage film. Because of this, some viewers thought that the film was a genuine documentary. The actor's parents received sympathy cards from people who believed they died during the events in the film.

Something similar happened with a 1990 documentary called In Search of the Edge. This short film details the life of the explorer, Andrea Barnes, who went missing when searching for the Earth's edge. The documentary illustrates many points to demonstrate that the Earth is irrefutably flat. Naturally, many Flat Earthers are big fans of this doc.

However, there is a problem. The documentary is fake. Since documentaries are expected to be a reliable source of information, this mockumentary was made for children to demonstrate how information shouldn't be accepted at face value just because it appears sincere. It was a test to provoke critical thinking... it wasn't meant to convince people that the Earth was flat! Yet there are people who defend In Search of the Edge, insisting that it is a genuine documentary.

Investigative journalist, Thomas Huchon, performed a similar experiment where he showed a fake documentary that "proves" the AIDS virus is man-made. Huchon has shown this documentary to students in over 80 schools because he believes "the best way to fight this kind of fake news discourse is not to give counterarguments, but to try to check the validity of the other person's argument" and "the only way to win is to teach young people how to think critically."

Jealousy is another big factor with conspiracy theorists. Every momentous occasion in the world needs to have an underlying conspiracy. Why did Celebrity A marry Celebrity B? Is it because they are in love? No. It's because they are in the Illuminati. Did man go to the Moon to see if they could achieve the unachievable? No, it was a scam to pocket billions of dollars.

In a 2017 study called Too Special to Be Duped, researchers noticed that people who saw themselves as special were more likely to believe in conspiracies. One researcher rationalised this observation by saying, "A small part in motivating the endorsement of irrational beliefs is the desire to stick out from the crowd."

Although astronomers are heavily educated, I have never seen one act like they are special while arguing to a Flat Earther. Astronomers tend to view themselves as logical, not special. Viewing yourself as special is unhelpful and potentially damaging.

Abraham Maslow famously put forward the Four Stages of Learning Theory. It breaks down our competence in skill, potential, and achievements in a certain activity into four stages. Nowadays, it is known as The Dunning Kruger Effect. It has four stages –

i) Unconscious Incompetence is when you are oblivious to how unskillful you are.

ii) Conscious Incompetence is when you are aware how bad you are at something.

iii) Conscious Competence is when you consciously exceed at an activity.

iv) Unconscious Competence is when you effortlessly perform to the best of your ability.

Many conspiracy theorists are trapped in the first phase – Unconscious Incompetence. No matter how many times their argument is proven to be outdated, sloppy, vague, inaccurate, or flat-out wrong, they won't change their mind. It's easy to get stuck in Phase 1 because no one wants to go to Phase 2. In Phase 2, you have to acknowledge that you are not good at something. But you need to push through it to get to Phase 3 and 4. If you can't acknowledge you can be wrong, I'm afraid you are in Phase 1.

There's a similar concept called The Bias Blind Spot. This is when you can't see your own hypocrisy. When this was tested in Stanford University, volunteers were asked how often they would be affected by different biases compare to normal people. Nearly everyone said they were among the few who never gave into their biases. But this can't be possible because the majority can't all think they are in the minority! Even when this fact was established to the volunteers, they firmly believed that everyone else must have overestimated themselves but they definitely didn't. This is why conspiracy theorists think they are the one who can see the truth while others miss it.

It's tempting to buy into a conspiracy. If there is a global cabal out to get you, you can blame "them" for your shortcomings. Many conspiracy theorists write hate-speech on Twitter. They accuse celebrities and political figures of being Freemasons, Illuminati, or even murderers. Unsurprisingly, these accounts are reported and suspended within a few days.

But guess what? According to the conspiracy theorists, they didn't get banned because they violated the rules of Twitter. These "Truthers" didn't get removed because they made racial slurs and homophobic remarks even though they signed a contract stating that they understand that sort of behavior was unacceptable when they joined the website. They got removed because of Them. Y'know. Them. The CIA. The FBI. The NSA. NASA. The New World Order.

Get fired at work? Your boss is probably working for Them.

Your friends not talking to you anymore? You must be getting close to the Truth.

The problem with losing friends when you become vocal about conspiracy theories is you start getting new friends that think the same way as you. When you are in a conspiracy community, you are in a group that listens to you. They encourage you. They believe you. No more rolling eyes or sniggers. When you explain your theories about how the world really works, you have their undivided attention.

The sad thing is that a conspiracy theorist may be so desperate to be believed, they will accept anyone, even if their other beliefs are challenged.

I have spoken to Flat Earthers that are homophobic, racist, transphobic, chauvinistic, and xenophobic more times than I count. Some Flat Earthers that I spoke to defended Adolf Hitler and claimed that he never killed anyone. The bizarre thing is that Flat Earthers who don't believe these things will defend Flat Earthers who do. They will turn a blind eye to defending a genocidal maniac so long as they believe in Flat Earth. When I pointed out how preposterous this was, they said, "We don't have to agree on everything." And that's true. They only have to agree that the Earth is flat.

So, what are you supposed to do when arguing with a Flat Earther? Astrophysicist, Brian Cox, said that they are so beyond reason, there is no point having a rational argument. In the words of political activist, Thomas Paine, "To argue with a person who has renounced the use of reason is like administering medicine to the dead." The best idea according to Cox is to mock them to highlight their silly beliefs in the hope that will discourage others from accepting the debunked claim that the world isn't spherical.

However, recent search suggests that criticizing or mocking one's beliefs is the worst way to change a conspiracy theorist's mind. That causes them to get defensive, solidifying the belief further. Every person has what is called a Coherence Checker which decides whether new information can be assimilated with one's existing beliefs. If a belief, no matter how absurd, is rooted into a person's brain, tearing it away can make the person fragile, which forces them to cling onto it.

So, what happens if you disprove a misconception that a Flat Earther has? Will that change their mind? Most certainly not. In 2005 and 2006, researchers at the University of Michigan did a series of studies that confirm that misinformed people rarely change their minds when exposed to corrected facts. Many people feel humiliated if they acknowledge that they fell for a hoax or a conspiracy and instead, reject facts that contradict their beliefs. The lead researcher, Brendan Nyhan, stated that a conspiracy theorist's beliefs become stronger when they are contradicted as a self-defense mechanism to avoid acknowledging their fallibility. They can't surrender the belief because they would have to surrender feeling special.

An Italian programmer called Alberto Brandolini devised a law that states, "The amount of energy needed to refute nonsense is an order of magnitude bigger than to produce it." It emphasizes why it is so difficult to debunk archaic or nonsense theories. Branolini's Law is also referred to as the Bullcrap Law.

This brings me to a concept called Cognitive Dissonance. This is when a person rejects evidence because it jars with a strong belief they have. A person can have multiple beliefs simultaneously, even if the beliefs conflict with each other. When I was a child, I believed that Adam and Eve lived in the Garden of Eden. At the same time, I was absolutely obsessed with dinosaurs. However, dinosaurs were around until 65 million years ago but the Garden of Eden (and the world) was created a few thousand years ago. So why did I believe both things even though they jar with each other? Because I wanted to believe them.

This is why people accept that crop circles were created by aliens even though Doug Bower and Dave Chorley admitted to creating the patterns (and then recreated them to prove it.)

This is why conspiracy theorists believe in Atlantis even though it was first mentioned by the fiction writer, Plato.

And this is why Flat Earthers believe in an archaic concept that has been disproven time and time again.

Can Flat Earthers Change Their Mind?

The purpose of this book isn't just to entertain the reader or to prevent them from embracing this outdated theory. I hoped that some Flat Earthers would read this book to look at both sides of the argument.

This brings me to the big question – Are there any Flat Earthers that stopped believing in Flat Earth?

Honestly, I struggled to find a single person who gave up on the FE theory. However, I did find a conspiracy theorist called John who strongly considered the concept. Luckily, John realised that Flat Earth was nonsense before he fully embraced it. John was kind enough to detail his experience. I feel this is worth including as it gives valuable insight into the mind of a conspiracy theorist.

This is John's story - "I was always big into conspiracies for two reasons. My dad was fascinated by aliens and the pyramids. The other reason was because I had a huge distrust for the government. What started out as a fascination on YouTube eventually turned to obsession. I watched documentary after documentary, whether it was about war, the banking system, the 9/11 attacks, Israel, Zionism, or the New World Order. Each conspiracy lead me to another one. Every truth felt like a puzzle piece and every one I found opened my eyes and pulled me in deeper and deeper. This journey lead me to conspiracies about the Moon landing and NASA. Without checking both sides of the argument, I assumed NASA was deceptive and corrupt. As I continued to research, I heard of Flat Earth for the first time. It seemed like every single documentary I watched from day one was to bring me to this Great Truth. I joined Twitter to find like-minded people to share my opinions. I always felt like I was a Truth-Seeker since I asked questions.

I was probably knee-deep in the FE conspiracy before I randomly came across people who argued against Flat Earthers. They are 100% responsible for stopping me believing in FE.

When this conspiracy was lifted from my mind, it caused a chain reaction. I realised not all the puzzle pieces matched like I thought. I realised that the YouTube videos I watched for five years were biased and I only looked at the clips that agreed with me.

I can proudly say that I am not a Flat Earther. I bought myself a telescope and now track and capture pictures of other planets and the International Space Station.

Most importantly, I made great friends with the anti-Flat Earth community. I also hope to find another Truth-Seeker like me; someone who is really looking for the truth without being blinded by bias."

www.ingramcontent.com/pod-product-compliance
Lightning Source LLC
Chambersburg PA
CBHW050403120526
44590CB00015B/1804